M000045772

In Global Warming We Trust:
Too Big to Fail

Stairway Press Revised and Expanded
Edition

Anthony J. Sadar

In Global Warming We Trust: Too Big to Fair

© 2016 Anthony J. Sadar All Rights Reserved

Print ISBN 978-1-941071-98-4
ebook ISBN 978-1-941071-99-1

This book is sold subject to the condition that it shall not, by way of trade
or otherwise, be lent, resold, hired out or otherwise circulated without
the publisher's prior consent in any form of binding or cover other than
that in which it is published and without a similar condition including this
condition being imposed on the subsequent purchaser.

STAIRWAY PRESS—SEATTLE

Cover Design by Chris Benson
www.BensonCreative.com

STAIRWAY⧋PRESS

www.StairwayPress.com
1500A East College Way #554
Mount Vernon, WA 98273 USA

Dedication

To my mother-in-law, Madge Clayton, and in memory of my father-in-law, Harry Clayton; and in memory of my parents, Anthony F. and Albina E. Sadar.

Foreword to Stairway Press Revised and Expanded Edition

THE CLIMATE HAS not changed much since the summer 2012 release of *In Global Warming We Trust: A Heretics Guide to Climate Science*—at least not much for the promoters of global climate doom. Yes, the disaster-monger tactics have changed somewhat, their hysteria has increased a bit, and much more money and politicking have been devoted to their dubious cause. The August 3, 2015 release of the Obama Administration's Clean Power Plan and the United Nation's late 2015 climate confab are grand cases in point. But, regardless of high-level machinations, the climate keeps operating as usual, changing in its substantially natural way.

No matter, trusting continues big time! After all, in essence what are far-ranging outlooks of global climate conditions—especially those fine-tuned for local areas? They are at best educated guesses by purportedly really smart people, and as such require trust by lesser entities, including other really smart people who don't have advanced degrees in climatology.

So, it comes down to trust, and the fact that people will believe what they want to believe, or are compelled to believe.

Yet, what if there are some really intelligent, independent, scientifically-minded folks out there with some impressive credentials, a lot of real world experience, and a dash of objective common sense, who question the ability of those other purportedly really smart people to be so utterly certain of their own prophetic powers? Would the input from the really intelligent, independent, scientifically-minded folks have any value in a free society, especially a society required to pay the bill for an enormously expensive gamble that the purportedly really smart people actually know what they're talking about?

And, we certainly *are* paying the bill. The federal government alone has poured billions of our tax dollars into research directed at substantiating preformed conclusions that humans are responsible for disastrous climate change and that increased carbon dioxide ("carbon pollution") produces only bad effects. Mounds of money are used to prop-up wind mills and solar collectors in the hope of averting an airy adversary in the form of increased severe weather events. In addition, a boat load of our cash floats into "education" of the public and students from grade school through graduate school on the culpability of people for climate catastrophe.

Like the giant financial institutions, the "climate-industrial complex"—as former U.S. Environmental Protection Agency senior analyst Alan Carlin and others have dubbed it—has now supposedly become "too big to fail."

But, are we investing wisely? Are there bigger issues out there in the real world that demand our serious financial attention and compassionate focus—issues that pose a bigger threat to humans and the ecosystem than some potential uptick in temperature levels. Two big threats topping the list are terrorism and abject poverty, both quite destructive to people and the planet, both within the means of our nation to greatly alleviate.

I am just one professional of the likely thousands that work in the atmospheric-science and related fields every day that see tremendous distortion by the news media, environmentalists, politicians, and even governmental bodies like the U.N.'s Intergovernmental Panel on Climate Change of what is and is not known about the earth's climate. So, this revised and considerably expanded Stairway Press edition of *In Global Warming We Trust* is my continued challenge to the final-form science of contemporary climatology foisted on an unsuspecting public. Arguments and insights presented herein are once again advanced not as "just another partisan broadside," but as a continued plea for more open-mindedness and tolerance in a discipline that absolutely necessitates such conditions for its optimal performance.

The bottom line is this: Authentic science requires that observations match predictions and no amount of bluster or conceit will change that. When scientists have lost humility, they have lost the ability to do science.

Anthony J. Sadar
Certified Consulting Meteorologist

Preface: A Practical Perspective

ONE HUNDRED AND sixty miles above the Arctic Circle, on the shores of the icy Chukchi Sea, lies a lonely military outpost, one of the many forgotten sentinels protecting the United States. Here at Cape Lisburne, Alaska in the summer of 1977, I began my career in atmospheric science. RCA Service Company hired me a year after I graduated with a BS degree in meteorology to provide civilian observations for the Air Force Weather Service Agency station along the arctic shoreline, and to train non-military radar technicians to do the same.

The sky that summer was mostly cloudy, temperatures were quite cool, and rain and mist were frequent. Though it snowed on July 5th, on two days temperatures hit 70°F.

On this small military base I experienced wilderness life. For a weather observer who grew up in Pittsburgh, Pennsylvania, daily life in the wilds of Alaska was anything but dreary. That summer, I spotted numerous species of birds and small mammals, and brown bears and caribou frequently appeared on the plains near the base.

One day dozens of magnificent caribou were grazing contently in the lush tundra not far from the majestic mount upon which the base's radar dome and meteorological tower were installed. In the less pastoral setting just outside the compound's kitchen, bears of various sizes tried their luck at dumpster diving. On one occasion, a small bear wandered into the main building of the base and had to be carefully ushered outside before it was missed and rescued by a protective, angry mother bear.

An undeniable treat for any weather observer was witnessing the aurora borealis. Cape Lisburne happens to be located within the Northern Hemisphere's maximum auroral frequency zone. On most dark nights the aurora was visible. (I say *dark* nights because in the summer the area above the Arctic Circle is known as the "land of the midnight sun"—the sun dips close to the horizon, but never actually sets.) There were times when the entire sky, from horizon to horizon, was filled with an awesome light show—a palette of mostly fluorescent lime green hues, gently swaying and swirling.

Several weeks before departing Cape Lisburne, I experienced the kind of anomaly only encountered in the field. The wind direction and speed sensor located on the adjacent hilltop weather tower ceased operating. Upon inspection, it was discovered that the four-foot long, bright orange and black instrument—a propeller weathervane that looks like an airplane without wings— had been mauled, most likely by a bear. I inspected the equipment and tower and ordered a new prop-vane.

A couple of weeks before heading back to western Pennsylvania, the new equipment arrived. Not wanting the new unit to be attacked again, I asked around the base for some ideas on how to dissuade curious critters. The cook recommended using garlic, because bears supposedly hate garlic. So before installing the new unit, I smeared it thoroughly with a homemade garlic paste. For my remaining time in the arctic chill, the equipment operated without interruption.

Later, working as an air pollution technician in the Midwest and subsequently in research and supervisory roles elsewhere, I again encountered interaction between the biosphere and the atmosphere, though mainly from birds nesting in and otherwise disturbing ambient air monitors. Each time effective deterrents had to be improvised to prevent equipment damage and data loss.

Though dealing with the encroachment of wildlife on meteorological instruments was challenging, the most important work that I've performed over my career has involved the public: assessing the impact, air quality-wise, on local communities from large and complex industrial operations, plant accidents, hazardous material incidents, and natural events. My classroom education complemented by in-depth field studies has enabled me to objectively and dispassionately evaluate the potential health hazards in such situations. And having a background in both theory and practice helped me when I needed to provide residents with an accurate and reliable, yet understandable and timely, investigative report. Numerous times fears from imagined harm were alleviated, while real problems were cooperatively addressed with effective action. Scientists can attract attention by playing to the public's fears, but they serve the public best when they identify and help implement solutions.

Overall, my experience with observations, data collection, incident evaluation, technical communication, and the general public is not that unusual. It is part of the vast knowledgebase assembled from the real-world experiences of thousands of atmospheric scientists above and below the Arctic Circle. I draw attention to this expertise not to put down others or invalidate the work they perform, such as theorizing and modeling, but to showcase a group of workaday scientists who appear to have been disenfranchised by contemporary climate science. For whatever reason, climate science is run almost exclusively by the officialdom of the academic community.

Experienced practitioners outside the academy have a lot to offer in terms of perspective, insight, tolerance, and just plain common sense. Indeed, these qualities, along with humility, are sorely needed in today's climate science. Theoretical and practical knowledge and skills must be better integrated to solve the most challenging environmental and societal issues the world is presently experiencing or will soon experience.

Introduction: What's Happening to Climate Science?

IT'S HOT, REALLY hot, and it is only going to get hotter!

The state of climate science tells us the world is indeed in trouble. We are facing a catastrophe of global proportions unlike any before in history. And it may already be too late to stop the unfolding cataclysm. Resources will be depleted. Governments will be stretched. Populations will be displaced. Our children's future will be limited.

All of this turmoil will only result if we don't effectively address the current state of climate science. The crisis in climate science isn't the result of a natural catastrophe. It's manmade: A cadre of scientific specialists has won the support of an army of career politicians, bureaucrats, environmental and social activists, academicians and educators, journalists, bloggers, technologists and consultants, and groupies of all stripes to spin a bit of understanding about the atmosphere into a trillion dollar bonanza.

Mega-computer power, sophisticated software, advanced numerics, and a whole lot of finesse have been mustered to pronounce unmitigated certainties about Earth's climatic future under increasing greenhouse gas concentrations.

Archimedes once said "Give me a long enough lever and a place to stand, and I will move the Earth." Well, the lever is certainly long enough—climate forecasts extend to the end of this century and beyond—and the new self-assured climate science professionals certainly have standing. So it looks like Archimedes was right because the globe is about to be moved—by an earth-shaking redistribution of power and wealth.

Unfortunately, those across the planet least able to care for themselves, the poor, will be shaken the most, followed by those without the financial savvy or eminent position to secure their hard-earned cash.

The poor will suffer from *imposed* resource depletion manifested by their dwindling ability to tap into earth's vast fossil fuel energy supplies, thanks to worldwide government regulations limiting energy generation to "eco-friendly" sources such as wind and solar power.[1] Governments will become bloated as more workers are hired to monitor compliance. Populations will be displaced as people are forced to relocate to either find jobs in the burgeoning (but iffy—think Solyndra and Beacon Power) "renewable" energy sector and government agencies, or to slum it out in low-rent districts after losing more traditional power company jobs. Our children's future will be limited as their ability to question pompous pedagogy or to challenge doctrinaire dogma in an attempt to exercise independence and creativity will be increasingly squelched by statist norms and standards.

And what is on the horizon for business and industry in the US if science continues to be held in the grips of a political ideology? People complain about business as usual, but wait until the crowd at the United Nations and like-minded Congressmen, academicians, and bureaucrats get their way. Because according to the UN, career politicians, and their friends, the verdict is already in: Human energy consumption, especially in the US, is disrupting the earth's climate and biosphere at large. So says the UN's Intergovernmental Panel on Climate Change (IPCC), so says the US Environmental Protection Agency (EPA), so say outspoken university researchers (who claim an indisputable consensus), so say the major environmental organizations and leaders of scientific societies, and so says a landmark decision by the U.S. Supreme Court in 2007 that was based on conclusions drawn by the IPCC. Next to have their say confirmed by the courts may be victims of hurricane Katrina, residents of Kivalina, Alaska (whose village is under siege from melting permafrost and sea ice), and who knows how many cities and states.

Is it really certain that humans are responsible for what some are calling "climate disruption"? Do government and academic scientists, lawyers, organized environmentalists, politicians, bureaucrats, and technocrats know enough to understand and engineer a fix for what they perceive as a broken climate system? Are draconian restrictions on energy use imposed by the President, Congress, and/or the EPA really necessary? In short, is all this expensive angst justified?

In a couple dozen brief chapters, we will examine the science and non-science driving concern over the greenhouse effect, anthropogenic global warming (AGW), climate change, global climate disruption, or whatever deceptive descriptors happen to be in vogue. After offering a succinct historical account of some of the major events leading up to the present hysteria, we will neatly address nearly every aspect of this often complex topic. What is the biggest climate driver? Is there really a consensus among atmospheric scientists on the future global climate conditions that are in store for us? Are greenhouse gas emission reductions, which are now mandated by the EPA for certain industries, really necessary?[2] Does "Climategate" matter? Why can't ordinary folks decide who and what to believe on this issue? What does progressivism have to do with climate change thinking? What roles do politics and funding streams play in climate science conclusions? What do some of the popular books on the topic have to say, and are they reasonable? All of these questions and more are tackled within these chapters.

You may find it profitable to read the book straight through. However, since each chapter (many containing pertinent sidebars) laconically addresses a different aspect of the climate change issue from my own observations, real-world experience, and considered opinion, the chapters can be read in just about any order the reader prefers.

In some ways, climate science is like a schoolyard with its in-crowd, geeks, bad guys, know-it-alls, etc. In this schoolyard of the supposedly settled science of climate change, I believe the public is not getting the full, true story about what is actually known and what are merely just-so stories and hand waving arguments. Many of those called childish names such as "denier" are not against popular and worthwhile ideas such as conservation, energy efficiency, common sense waste minimization, and so on. They are, however, opposed to bullies, questionable activities, and unfair games in which the rules constantly change. (And I'll have much more to say about this.)

Now it's time—especially with the 2016 elections looming—to bring some well-reasoned and measured insights to discussions about this critical issue, particularly since climate change confusion is driving calls for ominous and radical changes affecting our energy supply.

The key ideas that need to become part of the ongoing climate change discussion include:

The pursuit of science requires freedom—the freedom for scientists to use their individual skills, knowledge, and perspective in evaluating data and hypotheses.

Every scientist should be encouraged to practice humility. Unfortunately, arrogance abounds in climate science as demonstrated by: a) the presumed knowledge that humans are causing long-term, disastrous, global climate change and b) the absurd belief that we can confidently predict the future of the earth's climate out to the end of this century and beyond. As one author declares, "Humankind has the potential to alter the climate of the Earth for hundreds of thousands of years into the future. That I feel can be said fairly confidently."[3] You know you are dealing with arrogance when scientists resort to bullying and self-righteous arguments where they should welcome civil discussion and reasoned debate.

Progressive (leftist) ideology, the driving force behind mainstream environmentalism, has caused untold harm to environmental science. (Note that in this book I use the euphemisms "progressive" and "progressivism" to refer to political thinking and goals that should have no influence over scientific findings. Progressives range from people who are anti-industry statists to people who espouse radical ideologies such as Marxism. In short, leftist is a good descriptor and will frequently be used.)

Scientific illiteracy has enabled a great deal of deception regarding what is known and what can be predicted about climate change. Students and the public alike are doomed to simply trust, without question, climate science proclamations. Teaching "final-form science" (unassailable conclusions) and dogma as science have contributed to this illiteracy by short circuiting students' understanding of how science works (that theories, for instance, are built upon assumptions and have inherent limitations) and by discouraging students' own creativity and perhaps even their interest in pursuing science as a career.

A crisis-driven science is like a shady business. A supposedly authoritative consortium (such as the IPCC) identifies an urgent condition (anthropogenic global warming), financial backing is procured (largely from the deep pockets of Uncle Sam), solutions are proposed (altering lifestyles, shutting down coal-fired power plants), services are offered (education, research, consulting, trading for carbon credits, etc.), and oversight/enforcement is mandated (national and international bureaucracies). Everyone seems to be cashing in on the doomsday predictions, from private companies (consulting and technology firms) and academic institutions (university research and education) to governments with their expanding power and workforces. The big losers are, as usual, the ones stuck paying the bill—the middle class taxpayers—and the world's poor. Science also ends up losing thanks to a system of penalties and rewards favoring the crisis-mongers.

In sum, I was motivated to write this book because I have witnessed the increasing distortion of the science behind climate change and the overconfidence and arrogance behind long-range global warming predictions.

I also want to assure readers that I have no direct or indirect financial interest in Big Energy (or little energy, for that matter), though I must admit that I enjoy the comforts afforded by modern energy.[i] I do claim, however, to have a unique insider's perspective on how the atmospheric and environmental sciences operate today.

Meanwhile, back at the UN and in the halls of Congress, the ivory towers of academia, the offices of the EPA, and the courtrooms across America, based on the accepted consensus science and politics, the hue and cry continues to mount to save the Earth from manmade climate change. A headlong rush is underway to change the way we get and use energy. But this change will not be for the better. So before group-think and political mandates turn the lights out in America, let's see if a little common sense can be employed to avert a genuine manmade catastrophe.

---

[i] I must also confess to a cigar on occasion, but that's the only connection I have to Big Tobacco. And, oh yes, this past June, I bought a new car, a 2005 Saturn Vue, fully loaded (thanks Koch Brothers!).

Anthony J. Sadar

Climate Change, It's Personal

MANY KNOWLEDGEABLE SKEPTICS of the manmade climate change hypothesis lament the incessant ad hominem attacks rather than fruitful debate of this important societal issue.

Alan Carlin, the retired senior EPA analyst who had challenged the Obama administration's faulty climate science, in his new book *Environmentalism Gone Mad* (Stairway Press, 2015), noted that those pushing the "global warming doctrine" have almost always "refused to openly debate the scientific issues raised by skeptics but instead derided them or questioned their motives or sources of funding." This was witnessed in early 2015 with Arizona Rep. Raul Grijalva's attack on several prominent atmospheric scientists who dare to defy the authoritarian "consensus" on climate. These veteran scientists include MIT emeritus atmospheric-science professors Richard Lindzen, Georgia Institute of Technology earth and atmospheric science professor Judith Curry, and climatologist Roy Spencer.

Characterizing your formidable opponents as nut jobs, idiots, or shills is a technique for the lowest form of debate and the realm of spin-doctors, not for the honorable scientific profession. But mischaracterization is perhaps the best way to win an argument when the audience, in this case the general public, is ill equipped to understand the complexity of the topic. The public is destined to choose sides on the issue based on their trust in and likeability of the messenger. Hence, those hyping a disastrous climate future, with much help from the main stream media, will make themselves out to be reasonable, friendly, and trustworthy. Opponents of the dire futurists are then simply portrayed as untrustworthy fools, as is anyone who would believe the contrarians.

This disingenuous strategy is a big challenge for those of us who work daily in the field of science that strives to understand objective reality. A big part of that field is the application of what is commonly called the "scientific method" where the major components are observation, hypothesis, and testing. Once again, Dr. Carlin points this out in *Environmentalism Gone Mad* by stressing that the crucial "catastrophic anthropogenic global warming" hypothesis, which asserts that rising carbon dioxide concentrations will dramatically increase average global temperatures, "does not satisfy the scientific method" largely because observed reality has not matched predictions. Consider that, aside from the *one* surface temperature analysis recently released in the journal *Science,*[ii] numerous temperature measurements have revealed that the globe has experienced a relative flat-lining of temperatures for nearly two decades—this despite man's best efforts to stay alive and comfortable with carbon-based fuels.

Furthermore, proper scientific practice mandates that climate science conclusions should be based on the scientific method rather than consensus opinion. Such opinion is typically fostered by government largesse and groupthink that conforms to a particular ideology, leaving the resulting conclusions quite questionable.

What's happening in the field of climatology hits close to home. From my teenage years until now, my personal, academic,

---

[ii] See Karl, et al., June 26, 2015, *Possible Artifacts of Data Biases in the Recent Global Surface Warming Hiatus*

and professional life has been in the atmospheric sciences. Early in my career, I became interested in the greenhouse effect, then global warming, then climate change, now the carbon-pollution/extreme-weather issue. I have carefully followed this morphing monstrosity for thirty years. During that time, I have taught meteorology *and* climatology for Penn State and other institutions, and, as an air pollution meteorologist, have completed numerous air modeling projects. (Note that air pollution meteorology is a subset of the meteorological profession that includes climate-change studies.)

Regardless of my background and passion for the profession, I am branded with the profoundly nonsensical "climate denier" label because I remain unconvinced based on the lack of scientific evidence that humans have much meaningful (and certainly not disastrous) influence over the complex global climate system in the long run.

And thousands of seasoned practitioners like me, who have good reason to remain skeptical of man's damnable role in climate change, are targeted for derision by political opportunists, closed-minded, arrogant scientists, environmental zealots, professional spin doctors, and generally those who have just a superficial knowledge of how science is supposed to work (like insouciant journalists).

So what? Many, if not most, atmospheric scientists, economists, academics, and entrepreneurs making a living off the climate change issue would likely say "it's just business, it's not personal."

But to me, it's personal.

Most of my nearly 35 years of professional life has been involved with atmospheric modeling in one way or another. (Note that atmospheric modeling is the tool used to both develop future global climate scenarios and to panic the public on meteorological mayhem.)

I began my scientific career in meteorology in the late 70s.

Back then, calculating air quality impacts of air pollution sources, such as smoke stacks and vents, involved using a simple statistical calculator and some basic graphs derived from empirical studies—a rudimentary form of modeling.

Over the years, with more powerful computers and sophisticated graphics, air pollution meteorologists, like me, were able to analyze in more depth and with finer detail contaminant concentrations as they spread from their emission locations.

Today, air-quality models are coupled with some of the very same meteorological models used in climate studies. In this way, state-of-the-science estimates can be made to determine whether, for instance, a proposed industrial facility will contribute to unacceptable deterioration of air quality.

Air pollution models have long been used to evaluate just about any significant operation from the smallest chemical plant to the largest nuclear and coal-fired power plant. Furthermore, the models are useful in anticipating the consequences of mundane releases of contaminants to catastrophic outbursts from accidents or terrorist attacks that disperse gasses or particles like chlorine dioxide or anthrax.

What I and so many other air modelers have discovered is that, as impressive as modeling has become, model results beyond the immediate downwind distance of the pollution source and within a relatively brief amount of time, are not very reliable, despite the awesome computing power available today. We know that dependence on their output is quite limited and to extrapolate too far beyond the bounds of the model assumptions is foolhardy.

Compare the experience of thousands of non-academic air modelers with the largely academic and government climate modelers. Their combined efforts have produced impressive results in scope and scale, yet, like air pollution modeling, their model outputs still boil down to limited guesses.

A bit of understanding about the global atmosphere has been

spun into a trillion dollar bonanza by essentially PR supporters. But, if realism and humility about the limitations of climate modeling doesn't set in soon with enough scientists and those of the general public who care enough to pay attention, then more than our supposed climate future with be in dire straits.

Nutshell History of Climate Change Hysteria

CONGRESS FINDS ITSELF in a quagmire over energy legislation. The fate of the U.S. economy lies in the clutches of the EPA. A brief, review of the past 50 years of climate change debate should be helpful in understanding the present state of climate science.[4]

The supposed enlightenment represented by the campus teach-ins of the 1960s and early 1970s slowly invaded conventional college classrooms and the hippie mentality eventually became the hip academic norm. Traditional methods and standards of education were thrown out the window. However, excitement over newly relevant topics such as the planet's imminent collapse from too many people and too much ice quickly waned when population increases failed to yield global food fights and Mother Earth began to melt her once advancing ice caps.

Up until at least the mid-1970s, the frenzy to rescue the planet from industrial chemicals, especially pesticides like DDT, was fueled by Rachel Carson's influential but slanted book *Silent Spring*. This work, published in 1962, sparked the modern environmental movement, providing activists with both a laudable goal (saving the planet) and clear enemies (industry and modern society). To *Silent Spring* fans, it was obvious that the modern industrial society needed to be disarmed of its weapons (synthetic chemicals). Regardless of the fact that it is the negligent practices of industry and the wasteful excesses of society that should have been precisely targeted and not all of modernity, the battle to save the planet was on.

One battleground that soon became the main theater of the war was society's culpability for harmful climate change. But the early conflict was much different than it is today. In the 1970s, society was sensitized to two big issues: the Vietnam War and a worldwide cooling trend. In addition to cover stories in *Time*, *Newsweek*, and other popular magazines, the covers of books such as *The Cooling* by Lowell Ponte teased the public with alarming questions: "Has the next ice age already begun? Can we survive it?" Inside the book, Mr. Ponte notes "A handful of scientists denied evidence that Earth's climate was cooling until the 1970s, when bizarre weather throughout the world forced them to reconsider their views." Back then, you were a despicable "denier" if you didn't agree that the earth was threatened by *global cooling*. And has anyone noticed that other generations also believed they were uniquely plagued by bizarre weather resulting from modern advancements?[5]

The front cover of Claude Rose's *Our Changing Weather: Forecast of Disaster?* pondered, "Will our fuel run out? Will our food be destroyed? Will we freeze?" The back cover claimed, "Northern hemisphere temperatures have been falling steadily since the 1940s. Glaciers are advancing once again. Scientists no longer debate the coming of a new ice age: the question now is when?"

Were scientists in the 1970s completely wrong? If the answer is "Yes," then we are in big trouble because they are the ones who educated today's veteran scientists.

Kids were prepped for the coming catastrophe. A brief book by Henry Gilfond, called *The New Ice Age*, boldly displayed on its dust jacket three large thermometers in a row with ominously declining temperatures.

Society was alerted to the cooling danger from another quarter—though one with a more hopeful approach. A Christian tract by Walter Lang and Vic Lockman asked: *Need We Fear Another Ice Age?* The pamphlet discussed a solution that some technocrats were considering to stem the encroaching ice: apply coal dust to the surface of the ice so that it would absorb sunlight and melt.

And students were taught to accept the inevitable consequences. For example, some were told that in the future there might be polar bears roaming New York City. (That turned out to be true, but luckily they have been confined to the Central Park Zoo). Future atmospheric scientists were given a foundation of knowledge for understanding the coming ice age through meteorology lectures at The Pennsylvania State University.

As we know now, those fears of the past were unfounded. *Titanic* alarms were sounded: watch out for icebergs heading south! But they raised the alarms for the wrong reason. Back then, we were told that particles released into the atmosphere by industry would cause thermometers to plummet. But now we've finally got it straight. Gasses produced by human activity (primarily carbon dioxide) are going to cause thermometers to soar.

Okay, so there has been some disagreement about whether temperatures will decrease dangerously or increase dangerously. At least they are consistent about one point: something terrible is going to happen and humans are the culprits.

After a while, enthusiasm for global cooling cooled, which could only mean that scientists were warming to global warming. They just needed a little push, which Dr. James Hansen of NASA was kind enough to provide in his testimony before Congress on June 23, 1988. Dr. Hansen announced "the greenhouse effect is here and is affecting our climate now." With that statement, some additional theatrics (the hearing room was purposefully kept warm and humid)6 and the cooperation of the weather (the

14

eastern half of the US was experiencing an unusually hot and dry summer), Hothouse Earth hysteria was off and running.

The race was quickly joined by the UN Environment Programme and the World Meteorological Organization, spawning the Intergovernmental Panel on Climate Change (IPCC), an organization that has since produced an ongoing series of climate change reports known to some as the "Climate Bible."[7] This holy writ continues to fuel all manner of environmentalist and compliant media hyperbole on worldwide environmental conditions.[8]

In the late 1990s, the IPCC apparently decided to prop up the climate cataclysm storyline by taking more aggressive action. Direct temperature measurements from after 1900 were cut and pasted onto proxy records from before 1900 to produce the infamous "hockey stick" graph. Because there aren't enough records based on direct measurements from before 1900, scientists use preserved physical evidence such as tree rings, ice cores, and lake and ocean sediments to estimate temperatures for prior years. However, indirect measurements aren't as reliable as direct measurements. In any event, the hockey stick graph replaced the longstanding historic temperature graph in the IPCC's official global climate report for 2001. The original graph had clearly, but inconveniently, displayed a dramatic "medieval warm period" from about the 10th to 13th centuries AD and a "little ice age" from about the 17th century until the mid-1800s.

Meanwhile, the hockey stick graph was touted in Al Gore's highly political documentary film An Inconvenient Truth and it even replaced the previously accepted graph in a popular climatology textbook.

In other words, the frequently cited graph showing an alarming global warming trend is based on the arguable assumption that tree ring and other proxy data can be used to accurately determine global temperatures going back hundreds or even thousands of years. Plus, by washing out previous periods of dramatic warming and cooling the new graph ensures that practically any subsequent substantial warming or cooling trend is going to look ominous.

Sure enough, there has been a recent rise in temperatures. It began in the mid-1970s and is now leveling off. The upward trend was attributed to what scientists in the 1980s called the "greenhouse effect"—a term that very roughly describes how warming of the planet occurs. In the 1990s, it was more popularly referred to as "global warming," a phrase that suggests a loss of equilibrium. Finally, "climate change" became the preferred catchphrase, perhaps because it suggests a loss of equilibrium and hedges against any weather variability that might confuse an already perplexed public.

Fortunately, "all's well that ends well"—particularly when you are confident that things are not going to end well. Luckily for Mother Earth and her children, scientists in academia have been given sufficient government funds to thoroughly research the atmosphere, and they are so confident of their conclusions that they refuse to entertain alternative interpretations. And thanks to the miracle of modern science, climatologists can now not only predict the future with absolute certainty, they can pinpoint its causes.

So it is proclaimed that we are definitely headed for a climate calamity—propelled by greedy industrialists and a callous working class who have managed to overpower natural forces such as the sun. Now if they can just tell me which horse is going to win the big race at Churchill Downs...

There is, however, a glimmer of hope. If we can convince the wealthiest 1% to pay their fair share, we might be able to spend their share just in time to save the planet.

SideBar2a: Climate Change: "A Holiday Affair"

It's funny what you can learn historically about popular

climatology from watching an old, mildly-amusing, holiday film. The 1949 movie *Holiday Affair*, starring Janet Leigh (as the widow Connie Ennis, the love interest), Robert Mitchum (as Steve Mason, the interloper), and Wendell Corey (as Carl Mason, Connie's fiancé), has a scene early in the film when Carl is first introduced to Steve by Connie. It's an awkward moment. Both Carl and Steve are standing in Connie's living room, warming themselves in front of the fireplace. Their small-talk goes like this:

Carl: Looks as though we might have a white Christmas.

Steve: That's right. Never seems like Christmas unless it's white.

Carl: That's right. Though, we never seem to get the big snows we used to when we were kids.

Steve: That's right. Just comes down slush now.

Carl: That's right. Probably got something to do with the atomic bomb.

Steve: Hey, that's right!

So, it seems that for quite some time, the popular sense has been that the climate can be disrupted on a massive scale by certain human activity. Although this revealing chit-chat makes for convenient humorous dialog in this fantasy film, there's nothing funny about the official script in the real-world of atmospheric science—a script that demands everyone mouth the dubious storyline of catastrophic man-made climate change.

SideBar2b: Climate Change Hype Turns 25

As noted, June 23, 1988, Dr. James Hansen of NASA officially kicked off the man-made global warming scare by proclaiming before a Congressional hearing and the world that "the greenhouse effect is here and is affecting our climate now." [iii]

June 23, 2013 marked the tarnished silver anniversary of this historic global warming hysteria. Dr. Hansen's confidence, sincerity, and humble demeanor captured political and environmental opportunists in a big way in 1988, as now.

Here was (and still is) a man and a cause that could propel up-and-coming politicians to new heights and fill eco-activist coffers to overflowing.

But, from the beginning, the fix was in. Theatrics was to take center stage.

Former Colorado senator Tim Wirth and his staff left the hearing room windows open the steamy night before the proceedings to make sure the room's air-conditioners were chugging away during the momentous event.

Soon after hearings the U.N.'s IPCC took over, and the rest is history.

So, celebrations are in order. Time to commemorate:

Hundreds of regulations on the books and coming down the pike to spike industrial growth and economic advancement;

Thousands of hours of political grandstanding about how human exhausts have fueled funnel-cloud fury, hyped hurricane

---

[iii] In reality, Dr. Hansen's statement on the "greenhouse effect" was something of a ho-hummer. The "greenhouse effect" is an explanation of how the earth's atmosphere is transparent to certain wavelengths of insolation that impact the earth's surface, but absorbs and reradiates within the atmosphere outgoing energy from the earth; thus, maintaining biologically favorable temperatures across the globe. Furthermore, a greenhouse is not a very good analogy for how the earth's atmosphere operates to retain energy. The operation is much more complex and therefore, as one popular atmospheric science textbook (*Introduction to Atmospheric Physics* by Fleagle and Businger, 1963) recommended many decades ago, should be referred to as the "atmosphere effect." Besides, the most-significant climate regulator, by far, is water in all its phases, as vapor, liquid, and solid.]

histrionics, and otherwise monkeyed with the meteorological mechanisms;

Hundreds of thousands of tax-payer dollars shoveled into bottomless pits of "free" energy project subsidies, redundantly redundant Keystone XL and fracking environmental studies, vacuous eco-activist indoctrination curriculum, and crazy carbon-sequestration schemes; and,

Billions of people in poverty who could easily be lifted out by access to abundant, low-cost, readily-available fossil fuels.

Toot the horns. Toss the confetti. These are accomplishments only a socialist party could celebrate.

To be sure, real-world global temperatures were on the rise and climate modeling was improving to the point where consistency between that rise and the science that simulated it was good. Throughout the '90s, climate modelers were able to match and roughly predict global temperature changes using the assumption that carbon dioxide had a steroid-like impact on an otherwise feeble climate system.

Yet, reality has a way of humbling even the humble. For more than 15 years now, the global average temperature has been stubbornly uncooperative with the predictions of research climatologists and the wishes of environmental religionists.

Following is a graph recently prepared by state climatologist Dr. John Christy of the University of Alabama in Huntsville that displays temperature response to greenhouse gases. The plot indicates estimates and predictions of the IPCC climate models for the tropics against actual temperature measurements.

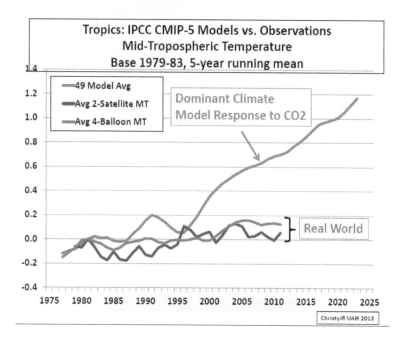

Source: Dr. John Christy, University of Alabama in Huntsville. From presentation given by Dr. Christy 30 May 2013 at a climate change conference in a technology park in Fairmont, W.Va, hosted by Rep. David McKinley, P.E. (R-W.Va.)

The top (red) line in the graph represents what the IPCC considers to be the best estimate of global warming. Such an estimate is being used to encourage scientists and leaders of authoritative bodies to promote dire consequences for the planet and its people.

But, the plot clearly shows that predictions are not successful in matching real-world global average temperatures traced by the two lower (blue and green) lines for satellite and weather-balloon measurements of the lower atmosphere. According to Dr. Christy, the disagreement of actual temperatures from predicted ones suggests that the models "are significantly wrong compared with the real world."

For the sake of humanity, hopefully the next quarter century will be marked by not only a continued level temperature trend but a more-even temperament regarding our ability to understand and predict the global climate decades ahead.

Such a temper would be a refreshing humility fitting to the scientific profession and worthy of a silver anniversary celebration.

A View of Atmospheric Science from the Field

THE HISTORICAL CLIMATE change furor has also created a crisis in atmospheric science. At issue is the way that science should be conducted. Under the current regime, certain ideas related to climate change are considered appropriate while others are considered inappropriate. And one group of practitioners is deciding which conclusions about the things that influence climate now and in the future are acceptable. Is science supposed to be this narrow-minded and controlled?

In many ways, today's atmospheric science practice is not that different from other challenging occupations.[9] It largely involves putting theory into practice in specializations such as weather observing and forecasting, climate research, and analysis of severe storms. My specialization is in air quality issues.

In the summer of 2011, I had the opportunity to attend a brief course in Boulder, Colorado on a state-of-the-science weather research and forecasting model. In fact, it's called the Weather Research & Forecasting (WRF) model. In addition to learning about this model, which is used in air pollution dispersion work as well as everyday forecasting and climate studies, I also had the opportunity to observe how young professionals from the US and around the world are applying their intellects and expressing their enthusiasm for the field. I was reminded of my early days in meteorology. Don't get me wrong: Even after thirty five years, I am still very excited about my vocation and enjoy the challenges it presents. Atmospheric science is, after all, a very engaging profession. But in Boulder that summer I found myself thinking about the trust and naiveté that novices bring to the field.

As young scientists focus on their particular roles, it's hard for them to grasp the discipline in its entirety, and they're not usually going to question the confidence their instructors or supervisors have in the validity of the computer models they use and the outcomes the models predict. The young scientists' primary purpose is to make contributions to the larger goal—not necessarily determine what the goal is or whether it is appropriate. For those who find employment in the climate modeling profession, it is reasonable to learn and embrace the reigning paradigm, but it may be difficult for them to see the paradigm's biases and limitations.

Among many tasks, young atmospheric scientists must learn to apply models properly. An atmospheric model such as WRF is like any scientific model in that it is a tool for understanding complex processes. Atmospheric models use complex algorithms and knowledge of chemistry and physics to simulate the operation of the atmosphere. For research and short-term forecasting, modeling tools such as the WRF program and coupled ocean-atmosphere global climate models are nearly indispensable. But their ability to produce accurate, reliable, long-range, worldwide climate predictions is questionable. Certainly, blind faith in that ability is unwarranted.

Scientists have done a good job over the years advancing atmospheric science and today it is a reasonably mature field. While atmospheric scientists have reason to be proud, they should remember that making scientific discoveries requires humility. After years of honing their skills and achieving modest goals, it's easy for scientists to fall into the know-it-all trap. They begin to create rather than discover truth; they believe that their simulations, particularly those employing high-speed computation, are reality; they believe that their models don't merely show what might occur under specific circumstances, they believe that their models show what *will* occur. Worse, they and like-minded colleagues inflate each other's egos until they soar into the stratosphere like weather balloons.

If scientists do not practice their professions with humility, then science is not being practiced. Though having general confidence in your abilities is fine, having confidence in your ability to know the future is not just unwise, it can be dangerous, particularly when others put their faith in such knowledge. Not only may young scientists be misguided, but activists may take advantage of the situation to promote a political agenda. (For example, intrusion and infusion of the leftist ideology into science may have done the most damage to climate science.)

In March 2012, I had the pleasure of attending the 10th Conference on Air Quality Modeling—a conference that the Clean Air Act requires be held every three years.[10] This public meeting is conducted by the EPA to help ensure that everyone with modeling expertise has an opportunity to offer their input to the government's modeling efforts and to discuss modeling methods and their limitations. Consequently, there is a community effort to identify the best air quality models for protecting public health and representing realistically the effects from a wide variety of pollution sources. And there was a spirit of mutual respect in evidence among the policymakers and the government, industry, academic, and independent model designers. That's a welcome departure from what's happening in climate studies, where progressive academics monopolize modeling efforts and have struck a very defensive pose vis-à-vis any challenges to their prescribed procedures and results.

The dominant view since the early 1980s has been that if the levels of specific atmospheric trace gases continue to increase, then it will lead to harmful climate conditions. This climate terror hypothesis originated among a relatively small group of scientists who had convinced themselves that the failure to reduce greenhouse gases would soon bring about a climatic apocalypse. They attracted a significant following among scientists and non-scientists alike. However, a crucial idea seems to have been lost during this period: the Earth's climate is a tremendously complex system.

In the 1980s, when climate change hysteria was in its infancy and conflicting perspectives were still permitted (and didn't provoke childish name-calling), the *Bulletin of the American Meteorological Society* published a compelling climate feedback diagram. This diagram, which should be revisited and contemplated by all climate modelers today, was explained by Alan Robock of the University of Maryland's Department of Meteorology in his article "An Updated Climate Feedback Diagram." [11] Under a line drawing depicting a tremendously complex network using interconnected arrows and labels, the author insisted that he "tried to make the diagram as simple as possible with as few arrow crossings as possible while still retaining the important climate relationships." In fact, the graphic was a reasonable illustration of the climate system—a system that one scientist aptly referred to at the time as a "horrendously complicated mess" during the Public Broadcasting System program titled "The Greenhouse Effect."

However, during the 1980s the atmosphere began to change in more ways than one, and that led to where we are today. Now groups claiming to represent the consensus among scientists are using sophisticated public relations tactics to convince the public that their assumptions, their models, and what the models say about the future are on-target and undeniable. As we will see later, organizations such as DeSmogBlog.com and the National Center for Science Education are waging aggressive public relations campaigns to ensure that only one perspective on global climate change is considered legitimate—the one that designates humans as culpable.

Given that the scientists claiming to represent the consensus view have assembled a vast army of non-scientists to promote and defend their beliefs, is there any place in atmospheric science practice for contrarians to be heard?[12] There are limited opportunities for dissenting voices to be heard in academia (such as being published in academic journals) and, sadly, industry and state and local government seem to follow rather than lead when attitudes change within the field. The scientists who are in the best position to challenge the dominant paradigm are scientists who work in a tangential field or have retired and are no longer practicing atmospheric scientists.

To be sure, there are scientists who work in (or have worked in) climate science who are capable of challenging the dominant perspective. The opinion piece that appeared in the January 27, 2012 edition of the *Wall Street Journal* and was signed by sixteen active and retired distinguished scientists is an excellent example.[13] This refreshing commentary presented the reasonable and defensible position that there is "no compelling scientific argument for drastic action to 'decarbonize' the world's economy."

There are more examples. Roy W. Spencer, principal research scientist at the University of Alabama in Huntsville and former NASA senior scientist for climate studies, presents his dissenting views in *The Great Global Warming Blunder: How Mother Nature Fooled the World's Top Climate Scientists*. Dr. Spencer reveals how climate scientists deceived themselves about the climate "forcings" and "feedbacks" that are explored via modeling.

Dr. Spencer explains in both quantitative and qualitative terms that "climate researchers have not accounted for clouds causing temperature change (forcing) when trying to estimate how much temperature change causes clouds to change (feedback)." He goes on to say that "they have mixed up cause and effect when analyzing cloud and temperature variations," and, "[a]s a result of this mix-up, the illusion of a sensitive climate system (positive feedbacks) emerges from their analysis."

In support of his hypothesis, Dr. Spencer examines recent satellite measurements of tropospheric air layers and concludes that "The climate system [is] relatively insensitive to warming influences." Dr. Spencer deduces from decades of Pacific Ocean temperature data coupled with satellite observations that "most of the global average temperature variability we have experienced in the last 100 years could have been caused by a natural fluctuation in cloud cover resulting from the Pacific Decadal Oscillation [PDO]...." (The PDO is a natural phenomenon that occurs primarily in the northern Pacific Ocean and that contributes to the flow of heat around the globe.)

Dr. Spencer and his colleagues offer compelling evidence that clouds play both forcing and feedback roles in climate change. Their work, however, is ignored or downplayed by the scientific and journalistic establishments with their leftist mindset that consistently views humans as a nuisance and ultimately bad for the planet.

Dr. Spencer concludes that "the 'scientific consensus' that global warming is caused by humans is little more than a statement of faith by the [IPCC]. There is evidence of natural climate change all around us if scientists would just take off their blinders."

It's going to take more than one person to get them to remove those blinders. More practicing, retired, and tangentially involved scientists must express their opinions about long-term global climate change and help liberate climate science from the grips of intolerance, political ideology, and other non-scientific influences.

In Global Warming We Trust

TODAY WE ARE told that within a few decades the globe will become intolerably warmer. The world as we know it will be drastically altered unless we act immediately to reverse our wayward lifestyles, particularly our wasteful energy practices.

Aren't we essentially being pressured to believe a long-range, global climate forecast? And isn't this pressure largely being applied by politicians and political organizations? Who among us would bet serious money on a weather prediction made a month in advance—let alone decades in advance? Yet the developed nations are under the gun to invest hundreds of billions of dollars in what amounts to a climate prophecy at a time when worldwide financial markets are already tottering. Doesn't President Barack Obama have enough deficit headaches without borrowing another trillion dollars?

Many people have had an epiphany much like the one that the late Michael Crichton had—that contemporary environmentalism, with its rigid tenets and unchallengeable proclamations, exhibits the worst characteristics of an organized religion. The environmental religion is headed by politicians (and former politicians) serving as its high priests and operating under a political cathedral (the IPCC). These adored patriarchs select the scientific data and climate conclusions used to write the sacred verses that are assembled to create the global warming scriptures. Their holy writ even includes a revised Book of Revelation with planetary disasters rivaling the New Testament's apocalyptic account.

Salvation is promised to those who redeem themselves from their carbonaceous sins by giving the priests control over their lives. Penance and indulgence take the form of "carbon offsets" which may be applied to offences including frivolous vacations, outdoor barbecues, the burning of incandescent bulbs, and driving Hummer SUVs (a truly mortal sin!).

Do not worry. There is also mercy in environmentalism. Offsets may be invoked by the inhabitants of industrialized nations to pursue economic expansion while enjoying their guilt-free contemporary lifestyles. Meanwhile, people living in third world countries are spared modern burdens such as dangerous power plants, dirty cement kilns, hazardous chemical factories, addictive pharmaceutical industries, sterile medical clinics, gluttonous harvests, and gushing purified water. Consequently, those with guilty consciences can vicariously enjoy the back-to-nature lifestyles of loin-clothed aboriginals foraging for food in lush rainforests to feed their gaunt families (provided they ignore the roughly one million people worldwide who die annually from malaria).[14]

How have we come to almost universally accept this new religion based on dubious prophecy that condemns so many poor souls to a living hell and limits the salvation offered by free economies? We have the global warming missionaries to thank for that. Better known as "teachers" and "professors," these missionaries travel to the furthest reaches of the education system to spread the depressing gospel that the Earth is endangered. With continuous indoctrination from grade school through graduate school, there are plenty of proselytes to be harvested.

It's no wonder that scientists succumbed so quickly and easily to the doomed-earth theory. The scientific community has been primed to accept calamitous forecasts about the atmosphere as if they had already occurred. First they were conditioned to believe that an extremely complex climate system is primarily controlled by a simple scant gas, carbon dioxide (CO_2)[15]—despite the fact that the single biggest climate regulator on earth is water.

The global atmospheric temperature is mainly controlled by the three forms of water: invisible vapor in the air, liquid in the oceans and the clouds, and solid ice crystals in snow cover and glaciers. And there are other factors beyond human control that could have tremendous influence over the earth's climate. Some atmospheric scientists and physicists have suggested that variations in solar radiation and cosmic rays are dominant factors.

So before we surrender to a catastrophic climate change scenario projected to occur decades from now, let's put it into perspective with the very real present day calamities of expanding terrorism, mass starvation, disease, ethnic cleansing, potential economic collapse, and the like. With these exceptionally serious challenges already at hand, and recognizing the enormous complexity of the earth-climate system plus the relative paucity of knowledge scientists have about the system's operation, we desperately need to restore some humility to environmental research and education. Politicians and scientists must acknowledge a basic tenet of science: theories are based on assumptions and have inherent limitations. The same is true for forecasts—only doubly so.

Like an organized religion, environmental science is going through a period of extreme intolerance. For example, former Vice President Al Gore wrote, "We must… stop tolerating the rejection and distortion of science. We must insist on an end to the cynical use of pseudo-studies known to be false for the purpose of intentionally clouding the public's ability to discern the truth."[16] Given that attitude, any attempt at reform is likely to get the would-be reformer burned at the stake (in a carbon-neutral fashion, of course) for environmental heresy.

Defending the Faith

Climate Cover-Up: The Crusade to Deny Global Warming by James Hoggan with Richard Littlemore unintentionally supports the contention that environmentalism is the world's fastest growing religion.[17] Like other religions, it uses crises, real or imagined, to win converts. After the threat of a coming ice age failed to gain traction in the 1970s, leftist environmentalists needed a new urgent cause, so they shamelessly began pitching its exact opposite, global warming, as the new impending crisis.

As former White House Chief of Staff Rahm Emanuel observed, "You never want a serious crisis to go to waste. And what I mean by that is an opportunity to do things you think you could not do before."[18]

However, if a crisis is an opportunity to get things done that otherwise might not be possible then leaders may be tempted to turn merely worrisome situations into full-blown crises.

And to get the most out of a crisis, it needs to be well managed. That requires good public relations. Enter the PR gurus and their consecrated corrective tome, *Climate Cover-Up*. Hoggan and Littlemore are public relations professionals—not scientists.[19]

Along with promoting their book, the authors operate a website, DeSmogBlog.com, that doesn't just defend the global warming faith, it launches a crusade.

The sinners are easy to spot in this religious narrative: Any individual or think tank supported by the usual industry suspects, "Big Oil," "Big Coal," "Big Tobacco," and other large, profitable businesses. In addition, anyone who dares to question Nobel laureate Al Gore, "the world's best climate scientists"[20] (the IPCC working group and task force members), and other such saints, or the anthropogenic global warming orthodoxy in general, can expect exposure, ridicule, and unrelenting belittlement by the ever-vigilant guardian angels. (In fact, a skeptic can even end up blacklisted in DeSmogBlog's "Research Database.")[21]

If just half of the alleged sins exposed by *Climate Cover-Up* turn out to have been committed (and I suspect more than half were), the contrarian camp has some serious penance to do. However, the authors would rather convince you that the contrarian position is indefensible and to accomplish that they are happy to use Orwellian newspeak, accuse contrarians of being in the pay of Big Energy, and deride contrarians as possessing dubious credentials. But such arguments cut both ways. Volumes have been written making similar accusations about anthropogenic global warming (AGW) proponents. Yet some human foibles, such as those evidenced in the "Climategate" scandal, are spun, minimized, and even defended by *Climate Cover-Up* and DeSmogBlog.

Let's talk about funding. *Climate Cover-Up* portrays Big Energy as inherently evil; therefore, anyone funded by Big Energy is tainted. However, Big Energy's contributions to AGW critics are quite small when compared with the largesse of Big Government, which *Climate Cover-Up* apparently favors, because it is behind most of the research that says AGW is real and all of the programs that force people to reduce their carbon emissions.

And what about the U.N.'s IPCC—is it committed to discovering and disseminating the truth? The organization defines its role as "...to assess on a comprehensive, objective, open and transparent basis the scientific, technical and socio-economic information relevant to understanding the scientific basis of the *risk of human-induced climate change, its potential impacts and options for adaptation and mitigation...*" [emphasis mine]. In other words, the IPCC's mission assumes from the get-go that AGW is a fact. In his movie *An Inconvenient Truth*, former Vice President Al Gore unleashed a quote from Upton Sinclair against his opponents: "It is difficult to get a man to understand something when his salary depends upon him not understanding it." But the very same quote could just as easily be applied to the IPCC.

All that aside, the authors of *Climate Cover-up* would probably agree that it's the facts that matter most. Here is where the authors, neither of whom has as much as an undergraduate degree in science, must take the veracity of manmade climate change on faith. They believe that faith is justified, because it is faith in "incredibly intelligent people who are doing a Nobel-prize winning job." The authors also rely on Al Gore, a politician who has learned quite a bit about climate science over the course of 30 years. However, does Gore's knowledge, acquired through political activism, trump a formal education and subsequent career in atmospheric science?

Nevertheless, it should only take a little science education plus a dose of common sense to see through the billowing incense when any individual or group, no matter how intelligent or sincere, claims to have deciphered the mysteries of the Earth's climate to the extent that they can confidently foretell global climate conditions 10, 20, or even 100 years from now.

I certainly agree with the authors of *Climate Cover-Up* when they say, "you should read up on climate science." In fact, I would start with a book that the authors misread and subsequently mischaracterized: *The Deniers,* by Lawrence Solomon.[22] The authors of *Climate Cover-Up* state that "Solomon admits [on page 45 of *The Deniers*] that none of his subjects were deniers. Not a single one." However, Mr. Solomon was clearly referring to the relatively small group of climate scientists featured in the early part of his book. His point was that several scientists were skeptical about the consensus climate science conclusions in their own specialties, but were otherwise convinced of manmade global warming. Therefore, they weren't true deniers. What this shows is that many people, even specialized climate researchers, are willing to rely on others' expertise. Mr. Solomon primarily focuses on true skeptics throughout the remainder of his work. Two dozen highly qualified scientists are conveniently dismissed by the authors of *Climate Cover-Up* by lumping them together with scientists who criticized certain aspects of the manmade global warming thesis but not its broader conclusions.

While *The Deniers* may be an antidote to some of the sanctimony exhibited by the authors of *Climate Cover-Up*, the website ICECAP.us does its part to counter the bunk dispensed at DeSmogBlog.com.

We can expect an even more aggressive (and at times bizarre) PR campaign against skeptics of the consensus view in the future. *Climate Cover-Up* was listed among the "recommended reading" in, and one of its authors contributed a sidebar to, a special cover story report in the May 15, 2010 issue of *New Scientist*. Under the title "State of Denial," the *New Scientist* asked, "From climate change to vaccines, evolution to flu, denialists are on the march. Why are so many people refusing to accept what the evidence is telling them?" Here was a respected science periodical suggesting that global warming skeptics are as foolish as those who continued to believe the earth is flat in the face of overwhelming evidence to the contrary, and in doing so are as ethically challenged as Holocaust deniers.

Skeptics of consensus climate science should be forewarned: In the eyes of the climate science establishment you are not only wrong, you are pathological.[23] And if you don't know it already, environmentalism does not take kindly to unbelievers.

SideBar4: 5 Facts about Climate-Change Science

As time rolls on, old challenges hang on.

One of the biggest challenges relates to the ambience of climate-change science.

So many folks are invested in an expectation of disastrous geophysical conditions resulting from modern lifestyles that are fueled by ancient energy sources. So, the big money (in the trillions of dollars) is on the continuation of supposed wacky weather hustled as proof of long-term, global climate change.

But, here are five reasons to remain unconvinced that humans are culpable for the naturally ever-changing climate.

1. Actual data trumps forecasts. No matter how you measure it, the global average temperature trend has flattened out over the past decade and a half. Even so, the tiny fraction of a degree increase in temperature for 2014 over previous years was hailed as being the highest on record. Though, regardless of this almost meaningless increase, there is a distinct possibility that temperatures will be once more dropping as solar activity and ocean circulations relentlessly work to redistribute heat across the globe.

2. By far, water is earth's big climate controller, not the carbon dioxide (artfully labeled "carbon pollution") associated with people living comfortably. Carbon dioxide makes up only 0.04% of the atmosphere. Compare this small percentage with water, which is a more substantial climate regulator. Water, in the vapor phase at 0% through 4% of atmospheric concentration and in the liquid and solid phases, is apparently the biggest climate controller on the planet.

3. The pernicious purveyors of long-term human-caused global climate change catastrophe constantly confuse weather with climate and cherry pick data to aggravate public angst over future meteorological mayhem. This permits, for example, huckster politicians to push the politically-laden baggage of the Social Cost of Carbon. The SCC ruse is not to be confused with beneficial balanced assessments like traditional Cost-Benefit Analysis. A thorough CBA would put some much-needed perspective back into any reasonable deliberation of climate concerns.

4. Then there is arrogance, the attitude quite at-home with promoters of climate-change hysteria. An attitude of arrogance gives away the game. No matter how brilliant climate

prognosticators are, nor how sophisticated their algorithms and super their computers, they are far from knowing with sufficient certainty the far future. Regardless, assuring highfalutin certainty seems to be the modus operandi for prophetic profiteers as they run their shakedown to save the planet.

5. Finally, real human misery requires immediate attention, and access to low-cost fuel goes a long way to alleviating suffering. At least a billion people don't have access to modern energy, living instead off truly dirty fuel sources such as smoky wood and dried dung. Yet, a mere one percent increase in so-called carbon pollution is estimated to satisfy the billion souls with the combustion of humane fossil fuels. As cheap, clean, abundant energy supplies power people out of poverty, high-cost, polluted policies potentiate and enrich politicians and their enablers. Thus, for ruling-class opportunists, the incentive is weak or nonexistent to make meaningful changes.

So, for the sake of the Earth and its inhabitants, when it comes to climate futures, we need to resolve to invest in less profit-driven science, and be more guarded and less gullible with diviners of disaster. More than a billion people will thank us.

The Command and Control of Climate Science

IN HIS JANUARY 20, 2015 State of the Union address, the president once again stressed urgency about climate change. And earlier that month, at the annual meeting of the American Meteorological Society (AMS) in Phoenix, Arizona, Environmental Protection Agency chief Gina McCarthy spoke with the same angst about the atmosphere.

Understandably, many have finally realized that the U.S. is truly in a world of trouble from too much pollution—not the "carbon" kind, rather the ideological kind.

Ms. McCarthy's January 2015 presentation to the AMS consisted of not only the typical derision of skeptics of manmade climate change and the distortion of climate reality, but included a rather delusional self-assessment.

Early in her talk, as usual, Ms. McCarthy denigrated any challengers to the so-called settled science of anthropogenic global warming as "flat earth" believers. She went on to claim that our biggest danger is in *not* taking action to stop an evident climate catastrophe.

Ms. McCarthy declared that "science is under attack as it has never been before" and elaborated that there is an "all-out attack on science in D.C. right now." Part of her solution is for scientists to be "more vocal," supposedly to help her in the fight to save the planet.

But for her part, Ms. McCarthy claimed that politics has "nothing...zero" to do with her assessment of the science behind climate change. However, her true assessment seems to range from zero to 100 percent political with reality being much closer to 100 percent political than zero percent. Much of her career has been closely tied to politics—apparently the kind aligned with statism—especially as an active state and federal bureaucrat.

Ms. McCarthy's position of authority and her enthusiastic personal commitment and demeanor demands our attention, or perhaps, even our subservience. Make no mistake, the Obama Administration is practically a driving force and potent fuel in climate science at the present time (e.g., the new Clean Power Plan). The feds set the tone and those who are still honestly unconvinced of a looming disaster (aka "deniers") will not be tolerated. On the other hand, a fountain of federal funding is flowing for projects to evaluate weather and climate data with respect to how it proves that humans are altering the atmosphere. And, the compliant expectation of continued global warming is still the modus operandi in the atmospheric science field, even though, in spite of confident climate outlooks and a slight increase in global average temperature in 2014, temperatures have essentially leveled off for more than a decade and a half.

So, studies in heat-related stress are in, studies in cold-related stress are out, regardless of the fact that fatalities from cold snaps can beat fatalities from hot spells by a wide margin.

There are many experienced atmospheric science practitioners like myself who have a different perspective and who represent no corporate interests and are *not* connected with fossil fuel industries. In my deliberations with numerous environmental professionals, so many have expressed some doubt (most much doubt) that humans are largely responsible for long term, disastrous, global climate change.

Yet, the marching orders from the president with his

administration's rhetoric and the new Climate Action Plan are to promote and finance dubious renewable energy and carbon sequestration projects while warring against evil, but reliable, abundant, cheap, poverty-alleviating, job-creating and job-sustaining, fossil fuels. Mother Earth must be defended at all costs...her children, not so much.

Forget the ethereal nature of long-range global climate predictions, the administration seems to have found a solid scary problem to hype, "solve," and leave as a legacy. Besides, Ms. McCarthy reminded the AMS audience that President Obama has claimed "climate change is a moral issue." Moral for sure, because unfortunately, if the administration's command-and-control of climate science practice persists, in years to come we'll discover too late that the legacy was one of expanding poverty, contracting liberty, and misdirecting science.

Sadly, the White House is encouraging a delusional and disastrous climatology. Their actions are reminiscent of a term in meteorology called the "triple point." The official *Glossary of Meteorology* (American Meteorological Society, 2000) gives one pertinent definition of the triple point as a "junction point within the tropics of three distinct air masses, considered to be an ideal point of origin for a tropical cyclone."

Beyond the atmospheric science profession, *politics*, *power*, and *profit* come together at a triple point to form the ideal perfect storm of destruction that has diverted billions of dollars from the fight against real, solid global issues such as terrorism and world poverty to less substantial issues such as human induced climate change.

Today, perhaps the best place to see the non-meteorological triple point in action is at the White House. Although President Obama's opening monologue at the February 2015 "Summit on Countering Violent Extremism" was, as expected, well-crafted and well delivered, his national security strategy released earlier that month warned that climate change is the real global challenge. The strategy states that climate change "is an urgent and growing threat to our national security, contributing to increased natural disasters, refugee flows, and conflicts over basic resources like food and water." The strategy assumes dubious relationships between human activity and atmospheric vagary.

Yet, in the real world, use of synonyms for "savage" and "evil" from Roget's thesaurus have been practically exhausted in describing the atrocities perpetrated by the Islamic State.

Understandably, Americans are becoming increasingly concerned about the mundane terrorist realities emerging in the here-and-now, not the presidential airy imaginings of the by-and-by.

And so, not surprisingly, in the public's mind, the present immolation of people takes precedence over the tenuous future warming (or is it now cooling) of the planet.

Regardless, in his January presidential address, Mr. Obama asserted that "American leadership drives international action"— saying this in relation to leadership on fighting climate change.After all, the administration tops its foreign-relations policy thrust with the need to lead the world to climate utopia, and this by going backward to the future with wind mills and sunbeam collectors rather than forward with nuclear energy and fracked fuels. At home, the administration seems more concerned with the climate-changing potential of the Keystone XL pipeline invading the country, rather than life-changing Islamic radicals.

Ostensibly, once climate change is wrestled away from human influence and handed back to Mother Nature for her control, then as corporation chairman Arthur Jensen enlightened Howard Beale in the 1976 Academy Award winning film *Network*, "our children...will live to see that perfect world in which there is no war and famine, oppression and brutality—one vast and ecumenical holding company, for whom all men will work to serve a common profit, in which all men will hold a share of stock, all necessities provided, all anxieties tranquilized, all boredom amused."

The fantasy world of the current administration seems well matched to the musings of Hollywood. But, as with *Network*'s uncanny prediction of what television was to become, the script also seems to have summarized well the fanciful expectation of the White House, as long as the malleable masses go along with their bureaucratic betters.

But, reality can produce a lot of unscripted spit-takes.

It turns out, as the real-world climate is demonstrating, human changes to the global environment are, in the long run, over the long term, inconsequential. Some changes are better, some worse, but overall the changes, especially to the long-range global climate, are turning out to be trivial.

More destructive to the environment and the people who inhabit it is the force of the arrogant triple point of politics, power, and profit, itself. This is especially true when this triple-point force naively believes it can overpower the singular real-life force of nature, in both its capricious climatological and "intensely wicked" human forms.

SideBar5: Three Reasons Obama Administration Almost Certainly Wrong on Climate Change

During the early months of 2015, except for a sidetrack to rush an Iranian nuclear deal that may truly produce global catastrophic results in the future, Obama administration officials continued to attack the thousands of honest, experienced scientists and engineers who remain unconvinced that human activities are responsible for cataclysmic global climate change. There is no conclusive proof that people are causing such serious atmospheric arrhythmia.

Assaults on knowledgeable, but incredulous, professionals have come through recent sanctimonious statements from the

Secretary of State John Kerry along with juvenile name-calling by other officials in authority—such as, EPA Administrator Gina McCarthy and the President himself—who should act with a bit more dignity, let alone intelligent circumspection.

Regardless, there are at least three big reasons why the Obama administration is almost certainly wrong about human culpability for catastrophic climate change.

1. The "hypotheses" that *humans are largely responsible for long-term, dire, global climate change*, has so far been shown to be remarkably false. Actual climate data suggest that the larger responsibility lies with nature itself. Unfortunately, the hypothesis was declared a proven "theory" much too prematurely by some so-called self-identified "consensus." Apparently, the consensus was based on mere opinion, founded on faith among like-believers. After all, anthropogenic climate change has, to date, become an unsubstantiated prophesy. Actual climate data for nearly two decades belie confident predictions of global warming from human "carbon pollution."

2. Money can't ultimately buy the truth; but *money can certainly distort the truth*. The U.S. Treasury has lots of cash to support research and programs that promote the man-is-the-enemy-of-climate hypothesis. The saying, "You get what you pay for," applies here. Unfortunately, it's the taxpayer who is footing the bill for political science masquerading as climate science. The discovery of truth suffers from an influx of government cash essentially earmarked for finding a big human footprint stomped in the global atmosphere—in a sense providing kickbacks to supporters of "correct" climate programs.

3. Perhaps the best reason to be skeptical of political grandstanding of certainty in science is that *the objectivity of science is destroyed by the subjectivity of arrogance*. Much worse than the

obvious ruse, "Trust us, we're politicians," is the more subtle ploy, "Trust us, we're scientists." President Dwight Eisenhower's farewell address to the nation over 50 years ago contained a warning that bears repeating:

> [P]ublic policy could itself become the captive of a scientific-technological elite.

No one group of people, no matter how prescient they think they are, can trump the combined intelligence of the community at large. In this case that community includes the multitude of reasonable skeptics in the atmospheric science and environmental engineering fields who question the "settled" dogma of human culpability to a troubled global atmosphere.

For example, a counter to the oft cited, but quite distorted claim that 97% of climate scientists are true believers in manmade global warming, a 2012 poll of the members of the American Meteorological Society revealed a great deal of skepticism among its membership. Only about 53% of the respondents agreed with the assertion that people are primarily responsible for the recent global warming. And, even for respondents assuming the existence of increasing average planetary temperature, less than 40% of such respondents claimed that the global warming will be "very harmful."

But, regardless of the furor and fury from those in the bully pulpit who use unsettled science to advance political causes, humble open-mindedness and inclusivity may yet help solve some of today's truly desperate environmental and societal challenges.

More Command and Control Climate Science

IN LATE JUNE 2014, I attended the annual international conference of a prestigious environmental organization, the Air & Waste Management Association (AWMA), and presented a paper on atmospheric modeling.

To kick off the environmental conference in warm Long Beach, California, Ms. Janet McCabe, Acting Assistant Administrator for the Environmental Protection Agency's Office of Air and Radiation, gave the lead keynote address on June 24th. Regarding the previous day's U. S. Supreme Court decision on the EPA's control of "carbon pollution," Ms. McCabe said her agency was "very pleased with the decision." The high court's ruling allows the EPA to impose greenhouse gas emission limits on major existing industrial sources, and then to go after new sources with stringent restrictions. So, you know all is not well for certain industries (like fossil-fuel fired power plants) going forward.

Also, when asked for his recommendations to young professionals, Dr. Barry Wallerstein, the Executive Officer of the South Coast Air Quality Management District and keynote speaker for the second day of the conference, said he would focus on climate change and sustainability. In addition, he predicted no more nuclear power plants in California's energy future.

During the week-long affair, attended by well-over 1000 professionals, I had a chance to reflect on why an association of well-intentioned practitioners has so readily succumbed to what I believe to be a biased, wrong-headed conclusion that humans are responsible for long-term, global climate change. After all, the global temperature trend has been rather flat after more than 15 years, even though practically all climate models predicted rising thermometers.

For the most part, graduates in fields related to the environment (such as environmental engineering and science, biology, meteorology, ecology, and the like) have been well trained in the fundamentals of each disciple, yet at the same time have been somewhat indoctrinated in a perspective that is "progressive" (i.e., leftist, including socialism and statism). This seems to be especially true of the younger generation of graduates. Perception from leftist groupthink, a hallmark of the hallowed halls, likely makes a lasting impression.

To confuse matters, academic, government, and business progressives convolute some of the principles and language and techniques of capitalism to promote their programs—programs that would be hard-pressed to survive *without* the assistance of capitalism. Enticing concepts of entrepreneurship and profiting from energy efficiency are proffered as if they are the proprietorship of progressivism.

Furthermore, besides the apparent indoctrination from grade school through graduate school that has inculcated the "incontrovertible conclusion" that people are ruthlessly destroying the planet, man-made climate-change hype via the media and PR spin doctors has infused acceptance of boundless human culpability into the psyche of everyone from the general public to atmospheric-science practitioners.

In addition, along with continued onerous, complex federal regulations, there are literally billions of dollars available for researchers securing grant money, consultants advising on carbon credits, and technocrats proposing carbon dioxide control and sequestration contraptions. It's easy to cash in on assessing "the risk of human-induced climate change, its potential impacts and options for adaptation and mitigation," as stated in the role of the U.N.'s Intergovenmental Panel on Climate Change.

What's more, atmospheric prognosticators who sincerely believe they can see the far-off future of the global climate are sincerely committed to the cause. Sincerity and commitment can be admirable, but not even the smartest among us can predict the climate decades ahead with any meaningful degree of accuracy—as already demonstrated by the recent level temperature trend.

There is hope for intelligent diversity, however, as many professionals with a valuable alternative view are speaking up. In August 2014, just after the AWMA meeting, a conference was held in Las Vegas of those who are unconvinced of many of the claims made about human influence on climate change. This was the *9th International Conference on Climate Change*. (I unfortunately did not attend the noble gathering, and you can bet that Janet McCabe didn't either.) Related to the conference is the Nongovernmental International Panel on Climate Change (NIPCC), "an international panel of nongovernment scientists and scholars who have come together to present a comprehensive, authoritative, and realistic assessment of the science and economics of global warming."

Besides the thousands of knowledgeable scientists, engineers, and economists participating in the Las Vegas conference, NIPCC, and numerous similar groups, there are so many of us unaffiliated professionals who sympathize with the cause of returning increased perspective and integrity to an honorable profession.

With persistent broadminded scientific practice and the continued unfolding of climate conditions in ways not predicted by vaunted climate models, the future looks warm indeed for a turn-around in climate-science thinking.

Yet the challenge is formidable as the current government administration from the very top on down through key officials in the EPA work tirelessly to impose their ideology-driven science on the populace. "I'm from the government and I'm here to help" used to be the perennial joke among those who had to deal with government "helpers." Now the saying is truly laughable.

These days, not only is the government not here to help, the government increasingly appears to be here to harm.

At the federal level that has become obvious with the new healthcare law. And now the results from the administration's new Clean Power Plan and presidential Executive Orders on the prevention of global natural climate change by controlling some local manmade "carbon pollution" is doomed to follow suit.

It seems that politicians at the highest levels and their political appointees have hitched their own arrogance and sense of superiority to that of arrogant scientists who "know" the future of the globe's climate. The scientists know that the addition of paltry carbon dioxide to the atmosphere from the continued use of fossil fuels will cause the globe to swelter, while the politicians know a fast-track to power when they see it. Together they have built a band wagon that rolls over any opposition to a politics-science conflation. Resistance is futile. The wheels of the wagon go round and round with a political agenda that supports the "science," while such science supports the political agenda.

About the worst thing to happen to science is its control by politics. Yet such control rolls on in particular with respect to fossil fuel use and climate-change science.

Secretary of the Interior, Sally Jewell, impudently stated in August 2013 that she didn't want any "climate-change deniers in my department."

The administrator of EPA's office in Dallas said in 2010 that his way of taking care of noncompliant oil and gas companies was "like when the Romans conquered the villages in the Mediterranean. They'd go into little villages in Turkish towns and they'd find the first five guys they saw and crucify them."

Gina McCarthy, head of the EPA, speaking at Harvard Law School in July 2013 made it clear that, regarding the President's climate-rule edict to state and local air agencies and its imposition on industry, "We have no choice. That's what the president said. He's my boss and you're going to have to live with it."

As John Hayward observed in *Human Events* in October 2013, "The imperial bureaucracy serves the King alone."

It doesn't have to be this way.

Much of my many years of experience in the atmospheric science and education profession has been as a public servant. I have worked with, and continue to work with, many at the local, state, and even federal level who have an attitude of true public service.

But with the impetus from on high to follow the leader rather than serve the public, more enslavement and impoverishment can be expected as demonized cheap energy falls to bureaucratic necessity.

In his first inaugural address, Abraham Lincoln referred to the American people as his "rightful masters" while Ronald Reagan stated in his first inaugural:

> *It is not my intention to do away with government. It is rather to make it work—work with us, not over us; stand by our side, not ride on our back. Government can and must provide opportunity, not smother it; further productivity not stifle it.*

Good solution.

So, "yes we can" get back to the days when "I'm from the government and I'm here to help" could just possibly be true.

Anthony J. Sadar

SideBar6: EPA Budget Focus: GHG Trumps More-Serious Environmental Issues

One of the best ways to keep tabs on what's happening with air issues inside the Environmental Protection Agency (EPA) is through a newsletter called the *Clean Air Report* of the aptly titled online news service *InsideEPA.com*. The *Clean Air Report* is a (rather pricey) bi-weekly publication produced by Inside Washington Publishers, which, according to their website, "for over 25 years has provided exclusive, relevant news about the federal policymaking process to professionals who have a need to know about the process."

The March 27, 2014 issue of the *Clean Air Report* newsletter contains back-to-back budget articles that starkly reveal one of the major problems with the federal government's hype of climate change. The first article, "EPA FY15 Proposal Boosts Funds for States to Implement Climate Rules," begins by stating that:

> *President Obama's fiscal year 2015 budget proposal for EPA would significantly boost funding for climate programs including states' efforts to implement the agency's greenhouse gas (GHG) regulatory agenda, which may help alleviate states' fears of resource burdens in crafting plans to comply with EPA's GHG rule for existing utilities. ...*

The article that immediately follows this lengthy climate-rules budget article is "EPA Budget Proposal Leaves Fate of Key Air Toxics Assessment in Doubt." This second article starts with:

> *EPA's fiscal year 2015 budget proposal leaves the fate of the agency's major nationwide assessment of risks from air toxics in doubt due to uncertainty over funding for the*

program, which could potentially mean another delay for EPA's latest version of the assessment after scrapping a planned update last year due to resource constraints...

As an air-pollution meteorologist and educator with 35 years of experience, I can attest the fact that air toxics (hazardous air pollutants) exposure has more immediate and potentially more long-term impact on human health and the environment than the ostensive enemy-of-everything-good greenhouse gases. As the focus and public monies are directed by this Administration to GHGs, much-needed attention and funding for other serious environmental concerns, besides hazardous air pollutants exposure (like water infrastructure), will also likely suffer.

The passages above represent one comparison among so very many others that showcase misplaced priorities.

Real public health is not well served by pursuit of green hobgoblins.

Earth Day: Politicking Science

THERE ARE MANY people like me—people who have long had an interest in the environment. On the very first Earth Day, April 22, 1970, I biked to school with a sign proclaiming the day, and later I attended outdoor environmental forums at Carnegie Mellon University in Pittsburgh. I have participated in both paid and volunteer environmental projects throughout my career. My interest in the environment is serious, but it is not based on unreasonable concerns.

The science profession, like most honorable pursuits, has its opinion leaders, its majority opinions, and its minority opinions. However, when pure science is mixed with politics—a mixture that is sometimes necessary from a practical standpoint—there are obvious pitfalls that scientists should watch out for and try to avoid. The science profession must not allow itself to be held captive by any political philosophy, party, or agenda.

Highly opinionated and domineering personalities, pretentious viewpoints, and even malevolent politics inevitably enter into the mix. Strong opinions and domineering individuals seem most in evidence around "Earth Day," a holiday created to promote awareness of and appreciation for the environment that's observed on April 22nd of every year. However, a day on which politicking dominates pure science is not something to celebrate.

Over the course of my career in atmospheric and environmental science, I have observed a dichotomy:

51

environmental sciences are practiced one way in the field and another in the halls of academia. Scientists and engineers who do hands-on work in the trenches, and must contend with real-world challenges on a daily basis, tend to be far more skeptical about the claim that human activity is changing the global climate—especially changing the climate in disastrous ways. Academicians, who primarily view the world from their computer screens, conduct limited field investigations, and engage in far-flung theoretical excursions, tend to be avid promoters of anthropogenic global warming (AGW). These academicians, whose financial wellbeing (as well as standing among students and the public) depends on the urgency of their topic, have a powerful incentive to focus on simple human causes and, therefore, simple human solutions to incredibly complex global climate challenges. This narrow focus limits their ability to discover new facts, explore other hypotheses, and propose alternative solutions for the benefit of society.

If academicians sincerely believe that human activity is causing global warming, then shouldn't they be identifying and evaluating different possible solutions? AGW proponents appear to be locked into one solution: reducing carbon emissions. However, physicist Freeman Dyson has suggested that biotechnology could be employed to reduce the amount of CO_2 in the atmosphere by half without sacrificing the enormous benefits of industrialization.[24]

Since most students gain only superficial knowledge of science from their primary school, secondary school, and college educations, they must simply trust the consensus among presumed experts about any modestly complex scientific matter once they become taxpayers and voters. For example, when the "lies, damn lies, and AGW statistics" are proffered as proof that the earth is warming due to the excesses of comfort-seeking humans and that it will continue to do so unless people quickly become less comfortable, the inadequately educated will feel that their only choice is to comply.

However, there are alternatives.

You don't need to be an expert to have an informed opinion. Nor are experts always right. Individuals can begin to educate themselves about global warming by taking a look at the latest global temperature trend data (through 2014). Earth Day would be a good time to start your education and the National Climatic Data Center website (www.ncdc.noaa.gov)[25] would be a good place to start. The data, for instance, shows a steady decline in the average level of the world's thermometers from 2005 through the end of 2008, followed by a rise in 2009. Be advised that AGW activists obsess over the fact that 2009 and 2014, for instance, were among the warmest years on record. However, if you start with the understanding that you are riding an amusement park roller coaster, you will realize that peaks (and troughs) are regular occurrences. Otherwise, the way that you see things depends on which side of one of many big hills you happen to be on.

And what about government policymakers—must they submit to the consensus among the presumed experts? Government agencies have enormous responsibilities and must consider the rights and needs of all citizens. Their technical and non-technical staff must select, implement, and oversee the best available solutions to real environmental problems. That's a monumental task, and by and large government agencies have done yeoman's work dealing with environmental issues. For example, government pollution control agents work with both the general public and the industries under their jurisdiction, enforce laws and regulations in ways that are reasonable, and overall perform quite admirably.

However, government agencies and individual agents can also become politicized. When this occurs, the will of the people may be usurped by a political program or private agenda pretending to serve the common good. For example, the public may be told that the Earth's climate is very sick and time is fast running out, that the cure is going to be very expensive, and that we have no choice but to undergo therapy. But if the Earth and all of humanity have survived previous warming and cooling periods, as most climate scientists would surely acknowledge, then the situation is probably not quite so dire.

In 1968, just a couple of years before the first Earth Day (April 22, 1970), Stanford University Professor Dr. Paul Ehrlich predicted in his book *The Population Bomb* that if world population continued to increase at a high rate it would soon (within one or two decades) lead to the collapse of economic and social systems. In late 2009, just a few months before the 40th Earth Day, Dr. James Hansen warned in his book *Storms of My Grandchildren: The Truth About the Coming Climate Catastrophe and Our Last Chance to Save Humanity* that if the world continues to burn fossil fuels at a high rate "there may be a threat of collapse of economic and social systems."

Dr. Ehrlich's prophecy proved a total bust; despite more guarded wording, Dr. Hansen's will probably meet a similar fate. Both scientists seem to believe that human activity is self-destructive and outweighs any mitigating factors.

Dr. Hansen, the director of NASA's Goddard Institute for Space Studies, is probably the leading government science figure warning that human reliance on fossil fuel combustion is destroying the Earth. Dr. Hansen is also the veteran climate scientist behind many, if not most, of Al Gore's environmental claims.

In *Storms of My Grandchildren*, Dr. Hansen asserts that his thesis is founded not upon climate models but simply empirical evidence. This evidence is derived primarily from ice core samples used to estimate greenhouse gas concentrations and corresponding temperatures going back hundreds of thousands of years. Hansen also uses the extent of the ice sheet cover to estimate past climate conditions.

Because changes in these two climate "forcing" mechanisms seem to correlate well with global temperatures going back eons, there is apparently no need to seriously consider other causes. In fact, reasonable candidates for climate change mechanisms, such as clouds and cosmic rays, are either ignored or ridiculed by the author. Regarding the possible effects of cosmic rays on climate (studied by Henrik Svensmark of the Center for Sun-Climate Research at the Danish National Space Institute), Dr. Hansen simply dismisses the carefully documented proposal as "an almost Rube Goldberg concoction." In fact, while Dr. Hansen respectfully calls those who agree with him "scientists" he refers to those who disagree with him (regardless of their credentials) as "contrarians."

Much of the support for the conclusion that greenhouse gases (particularly carbon dioxide) drive current and future climate changes comes from climate models. A model is a tentative representation (typically computer generated) of current or future conditions based on an interpretation of how climates operate. However, there are two major challenges in developing and using climate models. First, there has to be a proper understanding of what the data says (and does not say) about climatic processes. Second, given that the Earth's climate is extraordinarily complex, the data should come from many, diverse climate observation points from across the globe over a long period of time. Even Dr. Hansen agrees that the present number and variety of climate measurements could be inadequate for the purpose of forming a clear and complete picture of the Earth's atmosphere.

Regardless of the lack of sufficient climate data, Dr. Hansen puts out a call to action in *Storms of My Grandchildren* based on his convictions: the US must drastically curtail its use of fossil fuels. Dr. Hansen is apparently so confident of his ability to see "tragic certainty," he does not hesitate to suggest that fossil fuels, particularly coal, should be eliminated as soon as possible. And so would many other people who have placed their faith in his interpretive and predictive powers.

However, there is good reason to question the thinking of a

man who makes statements such as: "The leaders Obama appointed in science and energy are the most knowledgeable people in the field..." and "The present situation is analogous to that faced by Lincoln with slavery and Churchill with Nazism—the time for compromises and appeasement is over." Tell that to the Third World children and grandchildren who could be spared so much disease and hardship with the help of inexpensive power supplied through abundant and inexpensive fossil fuels.

There is much in *Storms of My Grandchildren* that conservatives can agree with. Dr. Hansen breaks ranks with many AGW proponents and says that cap-and-trade is not a good way to limit carbon emissions. He encourages greater energy efficiency. And Dr. Hansen even advocates developing and building fourth-generation nuclear power plants. But conservatives will agree most heartily with Dr. Hansen for this reason: When he discovered that anti-nuclear activists used deception in attempting to discredit nuclear power, he exclaimed, "That's what began to make me a bit angry. Do these people have the right to, in effect, make a decision that may determine the fate of my grandchildren?"

That's a good point. Dr. Hansen should understand, therefore, how others feel about people who make decisions that could determine the fate of their progeny, and extend the same courtesy to global warming skeptics that he demands for fourth-generation nuclear power proponents.

My personal, professional, and academic experience tells me that there's much more to be learned about the hugely complex climate system. Simplistic, politically- or ideologically- motivated declarations of climate "facts" and "solutions" to uncertain problems will only distract us from gaining a more complete understanding of the atmosphere and the extent to which humanity can or cannot interfere with its natural operation.

We can make improvements to the environment everyday—not just Earth Day—if we accept the skepticism that is part and parcel of pure science, protect science from undue political influences, improve science education, and create and implement environmental regulations that are considerate of everyone's needs.

COPing with Real-World Catastrophes

People who grew up with computer games and now live in the world of computer-generated climate met in Paris starting November 30, 2015 for the United Nation's 21st annual Conference of Parties (COP21) to discuss strategies to defeat their ethereal enemy—manmade climate catastrophe. Meanwhile, those who think beyond keyboard theories and computer graphics engaged the war that continues to manifest itself in places like Paris, to save other innocents from imminent and future destruction by real world destroyers—Islamic terrorists. After all, global terrorism is a real and present danger that no Party in Paris or climate crusaderism will stop.

Stating the obvious to the oblivious: It's a pretty sure bet jihadists are not all that concerned about reducing their carbon footprint. They probably wouldn't even appreciate a planet-sensitive climate delegate inserting a daisy in their Kalashnikov.

Across the Atlantic, except for the Obama Administration and Democratic presidential candidates, Americans apparently realize the relative unimportance of manmade climate change. Concerns about such change languish at the bottom of lists of issues that really trouble the citizenry—lists with the economy and terrorism at the top.

The U.S. can certainly do its part to alleviate the destruction to people and the planet by ruthless invaders. Instead of wasting tax payer's money on *WarCraft* climate battles and sending delegates to COP gabfests, the President should be focusing the American people's hard-earned revenue and intelligence resources on the real global war—the one on terrorism.

Kids "Scared Straight" to Climate Activism

AS ATMOSPHERIC PHYSICIST and professor emeritus Garth Paltridge observed, "We live in an age where common sense and tolerance are supposed to be the basis of our system of education, but there is very little of common sense and absolutely nothing of tolerance in the public argument about the climate change business. Perhaps it is that people simply have a basic need for fairy tales and doomsday stories."[26] Or to quote the inimitable Steely Dan, "The things that pass for knowledge, I can't understand."[27]

Teenagers, and even many adults, are not too old to believe in fairy tales. But youngsters typically have more time and energy to devote to fantasies such as the global warming goblin. And adult activists know that children make persuasive protesters.

For example, from May 7 through May 14, 2011 kids the world over traipsed around in the "iMatter March" to convince adults that the most pressing global issue isn't gruesome terrorist attacks, abject poverty, or tyrannical governments. No, the biggest planetary peril is climate change.

The week before the marches, some of the participants were plaintiffs in a lawsuit filed in federal court against then EPA chief Lisa Jackson, Secretary of Defense Robert Gates, and others. The lawsuit claimed that teenagers have a "profound interest in ensuring our climate remains stable enough to ensure their right to a livable future." The plaintiffs' talking points came from the high-power climate prognosticator James Hansen of NASA, whose sincerity and ability to convince has never been in doubt. Additional support and encouragement for the worried young people was provided by numerous progressive organizations— organizations that have become exasperated by the apparently large number of unconcerned and lackadaisical adults.

Most adults have learned that one of the basic truths of life is that no one knows the future. So if someone claims to know that the so-called Rapture will occur on May 21, 2011, or the Earth will end on December 21, 2012, or the globe will be intolerably warmer in 2050, they're probably deluded, arrogant, or both.

More perceptive adults understand that the future is not fixed. Even when it appears that something bad is likely to happen, there is always the possibility that something good will unexpectedly intervene and prevent it.

Other adults simply don't care. After all, they have adult responsibilities.

So if the climate scaremongers can't terrify enough adults, the next logical step is to create fear among their children who will, in turn, pester the adults until they do something.

Oscar Wilde once said, "In America, the young are always ready to give those who are older than themselves the full benefits of their inexperience." [28] What was true in 1887 is just as true today, albeit in a slightly different form. [29] It's often those with few if any adult responsibilities who have the time and inclination to save the world. However, there is a good reason for not letting 13-year-olds drive automobiles, purchase alcohol, vote, or hold elected office: they lack the good judgment and maturity that come from experience.

Children are especially susceptible to outside influences such as manufactured entertainment (video games), manufactured celebrities (think of Justin Bieber and Khloe Kardashian), and now manufactured causes. Adults can look at the claims of the climate change activists and compare them to facts revealed by the

Climategate scandal, charges of bias leveled at studies containing political recommendations, and cartoonish distortions of reality presented in propaganda films such as *An Inconvenient Truth*.

Children, on the other hand, tend to accept adult authority (particularly the authority of teachers, the mass media, and people deemed experts) and don't yet have the confidence and skills needed to critically examine claims. Therefore, it is unfair and even unethical for adult activists to target children when the activists have been unable to convince more confident and discerning adults of a looming global catastrophe caused by otherwise beneficial human activity.

Children are indoctrinated from an early age by left-leaning educators to believe that the actions of those who oppose leftist ideology ultimately lead to sorrow and disaster. The children take these manufactured and packaged causes home to their parents— or in cases such as the iMatter March, children are encouraged to bypass their parents completely in order to more efficiently drive their message home. As Oscar Wilde continued, "Indeed, they spare no pains at all to bring up their parents properly and to give them a suitable, if somewhat late, education."

How much parents learn from their children and how much progressive educators contribute to progress are interesting questions, but one thing that's certain is that our education system is in need of improvement. A good first step is to encourage children to be independent, critical thinkers. Unfortunately, our education system has been politicized; children are often taught to be critical but not necessarily independent-minded. (In fact, they are often asked to participate in group exercises in which a few learn how to influence others, while the majority learns that conformity is the safest harbor.) Children need to learn to evaluate complex concepts, such as global climate change, free from intimidation by adults peddling big, scary, unsubstantiated scenarios and free from pressure to accept the lowest common denominator ideas of their peers. Otherwise, they will simply grow up to be trained promoters of consensus thinking.

Then again, isn't that the ultimate objective of the purveyors of the global warming goblin?

Why Can't Johnny Think for Himself?

IN THE SPRING, a young man's fancy lightly turns to... thoughts of life after graduation. But has the graduate received a quality education? Only time will tell. Above all, we hope that our graduates have learned to think for themselves. However, there are some doubts, because the answers to the age-old questions: "Why can't Johnny read?" "Why can't Johnny write?" and "Why can't Johnny do arithmetic?" don't bode well for the newest inquiry, "Why can't Johnny think for himself?"

If Johnny had a solid elementary and secondary school education in the "Three Rs" (with apologies to everyone who values correct spelling), today he might be more competent in applying the fourth R, *reasoning*. And we might not now be asking, "Why can't Johnny question and challenge claims about human contributions to global warming?" Or better yet, "Why does Johnny accept that the very same gas he exhales is going to destroy his world?" Has Johnny learned critical thinking—or has Johnny learned not to oppose consensus thinking? True enough, many well educated people believe that the same carbon dioxide (CO_2) they and all other mammals exhale is causing runaway global warming. And even more astounding, many of these same people believe we can calculate what this CO_2 will do to our planet at the end of this century. But that still does not explain why so many have caved in to simplistic and preposterous ideas or why the public fails to recognize the complexity of climate change and the highly speculative nature of long-term climate predictions.

Regardless of these failings, a rapidly growing number of atmospheric and environmental scientists and engineers are becoming anthropogenic global warming (AGW) skeptics. These scientists and engineers, who are usually not recipients of government cash allocated for AGW climate studies or are simply retired from the field, are discovering that there are substantive, alternative explanations to the "human-released CO_2 = AGW" formula. These explanations include the overarching, long-term natural balance between what is exhaled by people and generated by their industrial activities, on one side, and what is "inhaled" by vegetation and the oceans on the other side, plus the variation in solar radiation, the disproportionate impact of cosmic rays on cloud condensation nuclei formation at different altitudes of the troposphere, and the contributions from La Niña, the Pacific decadal oscillation, and the North Atlantic oscillation, to name a few important factors.

Why can't Johnny seriously consider differing views? The problem is that Johnny has not been encouraged to think for himself: to question others' assertions, to compare and contrast different views, to distinguish between an hypothesis and a fact, or simply to stand with a reasonable perspective in the face of opposing popular opinion. Instead, Johnny's independent research and thinking skills have been hobbled by an elementary and secondary education system that is more interested in manufacturing what it considers good citizens. And higher education generally does little to remedy the situation. By graduation time, Johnny has been more indoctrinated than educated. His cap and gown do little to cover his inability to understand cap and trade.

How is Johnny indoctrinated? Johnny is taught to accept the testimony of experts and the conclusions of academic studies. However, the words "expert" and "study" are often used to bamboozle people; experts are not always right and studies are not always accurate, conclusive, and reliable. Consequently, Johnny has become vulnerable to intellectual bullying at precisely the time when we need him to use his cognitive skills to help solve problems and develop innovative solutions. Unfortunately, it seems that Big Education would rather that Johnny simply acquiesce to its elitist solutions for imagined problems in order to keep the research grants flowing. Short shrift is given to genuine global perils that require immediate attention such as unsustainable government budget deficits, extreme poverty, insufficient potable water, widening terrorism, and more.

AGW challengers such as Brian Sussman see more pernicious forces at work in contemporary education. In his book *Climategate: A Veteran Meteorologist Exposes the Global Warming Scam*, Sussman quotes Vladimir Lenin, the architect of Russia's communist revolution, boasting, "Give me four years to teach the children and the seed I have sown will never be uprooted." Mr. Sussman, a former television meteorologist, is now a popular talk show host on KSFO Radio in San Francisco.[30]

In his accessible and partly historical account, Mr. Sussman lands a decisive blow against the insidious socialist and Marxist ideas that have infiltrated the green movement and have become a roadblock to balanced, rational thinking in the US and elsewhere. It's an interesting coincidence that Earth Day, a holiday celebrated in public schools across America, is held on Lenin's birthday (April 22) and that the very first Earth Day was on what would have been Lenin's 100th birthday (1970).

Sussman ventures out into the trenches and talks to scientists working at the front lines of climate change research such as David Deming,[31] a scientist who has conducted proxy temperature studies and is familiar with how conclusions that don't agree with anthropogenic global warming (AGW) are repressed, and Fred Singer,[32] a pioneer of modern atmospheric science who sees global warming as "an excuse to cut down the use of energy."

Mr. Sussman delivers his most potent sword thrust when discussing the cap-and-trade scheme in a section titled "Carbon Rich." He describes how a select few will skim transaction fees from heavy trading in CO_2 credits at establishments such as the Chicago Climate Exchange.[33] Then he states sadly:

> *While perfectly legal because the law will allow it, as it always does when immorality is legislated, these greedy, conscienceless investors will rape and plunder our once-great nation with hearty political approval.*

Even James Hansen, who instigated the humans-are-destroying-the-planet-by-emitting-carbon-dioxide hysteria, sees the cap-and-trade scheme for controlling CO_2 as wasteful and an invitation to corruption.

As Brian Sussman observes, it all comes back to education. Teach children the same thing day after day and within a few years it becomes hardwired in their brains. If educators really want to encourage critical thinking, then they need to get back to teaching the basic skills and facts that provide a foundation for independent research and analysis. And they need to accept that only a free market for education, in which every family has a choice of schools, can guarantee the diversity of ideas and perspectives that lead to creativity and innovation.

Critically Thinking about Climate Change

GETTING STUDENTS TO "think critically" has been a serious effort by educators for quite some time. Of course time after time we've seen that in practice the critical thinking desired is the thinking that questions traditional, conservative positions. But, if critical thinking is honestly what instructors are striving for, why not expand student thinking by challenging students to apply the technique in new, practical ways?

As a life-long atmospheric and environmental scientist and long-time college-science educator, I am constantly bombarded with material from a variety of sources, including many environmental groups. Take, for instance, what can be labeled "sales" literature that I recently received from the Environmental Defense Fund (EDF). The mailing contained a small double-sided poster that was titled "EXTREME WEATHER: THE CONSEQUENCE OF CLIMATE CHANGE" on the one side and "TACKLING CLIMATE CHANGE" on the other side. I will focus only on the "extreme weather" side here as an example for effective pedagogy.

What if a teacher were to display the Extreme Weather poster in the classroom and ask students to carefully consider its contents? The poster contains 6 text boxes, each describing a consequence of climate change: Wildfires, Extreme Heat, Storms, Droughts, Flooding, and Swelling Oceans. Take the contents of the Storms box, for instance. It claims:

Scientists have warned that climate change could bring stronger, more destructive storms. Superstorm Sandy—the largest tropical storm on record—brought those predictions crashing down on the Eastern U.S. on October 29, 2012. Responsible for at least 147 fatalities, 8.5 million people without electricity and $50 billion in damages, Superstorm Sandy's reign of terror extended inland to the shores of Lake Michigan and northward to Nova Scotia, Canada. [Emphasis in original.]

Each sentence can be evaluated literarily and scientifically.

As literature, students could be challenged to examine the style, flow, and tone of the message. The highlighted first sentence could be assessed for its real substance: Who are these "scientists" who have such a dire warning? How many are we talking about, 2, 10, every scientist? Is the statement too nebulous to even have serious meaning, regardless of the one example of Sandy that follows? Furthermore, phrases like "crashing down" and "reign of terror" could be parsed for their effect on eliciting deep emotions and inciting readers to "doing something to save the planet."

From the science perspective, how is "stronger" and "more destructive" actually determined, including considering measurement techniques, availability of historic records, increased population and property development, and the like? Further, what is meant by "largest tropical storm on record"? In reality, how extensive and extreme was the storm's "reign of terror"? In Pittsburgh, for example, the storm's "fury" was relatively light with some high winds and precipitation. Sandy did become a Hurricane, a category 3 over Cuba, but only a category 1 (the lowest level) off the east coast of the U.S. Does the fact of this low designation give some scope to the storm's overall intensity?

In addition, at the bottom of the poster we see the

claim: "Global Warming: *More Daily Record Highs in U.S. Than Record Lows.*" Starting (conveniently) with the 1950s and then jumping to 2009 through 2012, pie charts display proof of this claim. Here students can be encouraged to put statistical skills into play. How does the selection of data and time periods affect results and conclusions? Is the fact that the contiguous U.S. is less than two percent of the earth's surface important to consider? And, more generally, how are statistics used to enlighten or darken reality?

These are but a few suggestions for use in critical thinking in the classroom. The danger in this poster-checking exercise, from a "progressive" educator's point-of-view, is that students who critically evaluate eco-activist pulp may end up not buying what the environmentalists are selling. And that kind of thinking truly is critical.

I recall decades ago, in elementary school, my teachers would encourage students to "put on your thinking cap" when they wanted the students to—as modern pedagogy would say—engage their cognitive skills.

Today, I suspect that many teachers likewise encourage their students. Let's hope so, because now, perhaps more than ever, novel, independent thinking has some high hurdles to clear. And, this is at a time when contemporary global challenges still require some novel, independent thinking.

Regardless of dubious climate issues addressed by the EDF poster, there are worldwide economic problems in dire need of quick, effective, compassionate solutions and typical high-priority environmental issues related to energy, such as mitigation of "traditional" pollution from the burning of fossil fuels.

Take my field of air quality. The contaminants that require attention are the ones that can demonstrably foul the planet near and far, now and later—pollutants like air toxins and six "criteria" pollutants labeled so by the federal Clean Air Act, specifically, particulate matter, sulfur dioxide, nitrogen dioxide, ozone, lead, and carbon *monoxide*—not carbon *dioxide*.

Carbon dioxide, a so-called "greenhouse gas" that occurs abundantly in nature and is essential for plant life, has recently been targeted for drastic reduction from industry smokestacks and vehicle tailpipes by the U.S. Environmental Protection Agency.

Part of the concern is that carbon dioxide and other greenhouse gases like methane allow short-wave energy coming in from the sun to reach and warm the earth's surface, but impede the escape of the returning long-wave energy emanating from the planet. Thus, energy that should have radiated to space is trapped (absorbed) and in turn warms the earth's atmosphere.

To many, like those from EDF, that brief explanation, coupled with temperature graphs of dramatic recent global warming and climate model results predicting a sweltering orb by century's end from continued greenhouse gas emissions, is enough to indoctrinate students from grade school to graduate school in the belief that humans are about to destroy the future if something is not done immediately to stop them. One promoter in fact claimed that "[h]umankind has the potential to alter the climate of the Earth for hundreds of thousands of years into the future." In *The Long Thaw: How Humans are Changing the Next 100,000 Thousand Years of Earth's Climate,*[iv] author David Archer continued with "[t]hat I feel can be said fairly confidently."

With such confidence, coupled with near total control of the education system, a leftist ideology permeates the developmental atmosphere of the modern classroom. And with it comes a squelching of ideas that may provide the ideal solutions to challenges like climate change.

In college, the views academics impress upon their students are all too frequently based on partisan progressive politics, radical professorial notions, or hypotheses masquerading as well-established theories (for example, catastrophic anthropogenic

---

[iv] Princeton University Press, 2009

global warming).

But, the apparent current trend of shielding students from concepts that leftists think are bad science—like the reasonable proposition that nature, not humans, is in charge of climate change—hinders students' development of critical thinking skills.

And yet everyone has been told that one of the esteemed goals of modern education is to develop critical thinking skills. So, if you're reading any literature supplied by an environmental organization (or even industrial group), put on your thinking cap and think again.

Note that excellent resources for teachers and students to access to effectively counter some of the "facts" of the Extreme Weather poster review in this chapter can be found at www.icecap.us and www.drroysprencer .com and, in particular, a presentation by Dr. John Christy of the University of Alabama in Huntsville.[v]

SideBar10: The Inculcation of Final-Form Climate-Change Science

As college semesters wind down to a close, long holiday and semester breaks can give educators a chance to ponder the dismal state of science literacy in the U.S. The sad decline in robust science education is certainly part of the problem and is perhaps most obvious in environmental science classrooms. Contributing to the problem is the skewed content in many college textbooks on the environment and ecology.

While a part-time college professor of the physical, environmental, and atmospheric sciences since 1986 and a practitioner in the field since the late 1970s, I have had the opportunity to review and use numerous popular textbooks.

---

[v] www.globalwarming.org/2013/05/31/john-christy-climate-change-overview-in-six-slides/

I was disappointed to read in one of the latest textbooks—*Essential Environment: The Science Behind the Stories*, 5th edition (2015) by Jay Withgott and Matthew Laposata—the following distorted statement about those of us who dare to challenge the current groupthink on climate change.

> *Public debate over climate change has been fanned by corporate interests, spokespeople from think tanks, and a handful of scientists funded by fossil fuel industries, all of whom have aimed to cast doubt on the scientific consensus. (page 322).*

This typical misrepresentation is found within Chapter 14 of the book, titled "Global Climate Change." The subsection of the chapter is labeled "Are we responsible for climate change?" and contains a mere six long sentences crafted to convince students that there are no real honest skeptics to the "consensus" view.

In truth, there are many of us who honestly represent no corporate interests, are not involved with think tanks, and have no connection to fossil fuel industries.

As observed elsewhere in this book, in my discussions with numerous environmental professionals, nearly all have expressed some doubt (most much doubt) that humans are largely responsible for long term, disastrous, global climate change. Yet textbook authors continue to push the idea on vulnerable students that the matter is an open-and-shut case.

A 2012 poll conducted by the George Mason University Center for Climate Change Communication quizzing members of the American Meteorological Society (including me) about confidence that we know humans are causing climate change revealed that the so-called consensus claim is quite dubious. Setting aside the fact that consensus-building is the bailiwick of politicians, not scientists, and even assuming the existence of substantial anthropogenic global warming, no more than 40% of

AMS respondents to the George Mason questionnaire claimed that such global warming is dangerous. The survey results further challenge the official proclamation of the AMS that asserts that humans are largely responsible for climate change.

Several years ago, when I was a student in a doctoral science education program, we were rightly instructed to shun "final-form science," that is, scientific conclusions that claim to be established beyond any reasonable doubt and are dictated to students as absolute truth. Nevertheless, the dogmatic climate-change statements of "consensus" and "settled" science being inculcated on unsuspecting undergraduate students are exactly that—readily challengeable and practically challenged science portrayed as established, indisputable fact. By practically-challenged science, I refer to the greater than 15-year "hiatus" in global warming that has been confounding the hypothesis of human greenhouse gas impact on global temperatures.

In addition, practically challenged includes the objective observations and perspective of real-world, field professionals who experience and research a complex climate system that is far from explicable by subjective academic climate models.

Scientific literacy will be better promoted by textbook authors providing a broader—and thus more accurate—perspective on critical contemporary issues such as global climate change. Any book that claims to tell the story of science behind the headlines yet obfuscates how real-world science operates does a disservice to both students and society.

Graduates, Welcome to Reality

IS THE FRESH crop of college graduates who enter the workaday world each spring adequately prepared for the road ahead? Were they encouraged and empowered to be creative thinkers ...or simply worker bees? Not everyone has to become a creative thinker—not everyone wants to become a creative thinker—but progress in the workplace and world-at-large requires at least a few sources of fresh ideas.

We may not be able to produce creative thinkers on demand as some educators like to imagine. But at least we can provide students with opportunities to explore and nurture their creativity. The last thing we should be doing is teaching students that being a good team player is the only option. Not every sport is a team sport.

College students are ostensibly exposed to the fundamentals of language and literature, music and math, philosophy and physics, science and society. Though the range of learning may be sufficient for a scholastic foundation, with the exception of students who participate in work-study programs, a college education tends to be all theory and no practice.

Compare the academic and workplace environments. It's not just the physical surroundings that are different. There are two very different mindsets at work. Academics tend to see the business world as obsessed with making money. Businesses are less impressed today with college educations than they were a couple of generations ago. They see students as steeped in untested theory with little knowledge or even insight about how to apply what they've learned to the outside world.

Unfortunately for students, the career professor with little or no real-world experience often develops narrow and slanted views and presents those views to students as obviously correct. Naturally, most students go along either because they accept what they are told on the professor's authority or want to maximize their chances of getting a good grade. Do such professors prepare students for what awaits them on the outside? Not if the views they impress upon their students are based on partisan politics, radical professorial notions, or flimsy hypotheses masquerading as well-established theories.

This lack of real-world experience explains much of the naïveté encountered in academia. Though it's not necessarily a critical problem; there is after all something to be said for a place where people are free to conceive and develop ideas. However, the combination of naïveté and arrogance can be disastrous. For example, about one million people die each year from malaria because some gullible and anthropologically insensitive intellectuals think DDT is a four-letter word and have successfully campaigned against this efficacious pesticide. Fortunately, "A number of countries are in the process of reintroducing DDT for malaria control."[34]

As Thomas Sowell observed, "What is most frightening about the political left is that it seems to have no sense of the tragedy of the human condition. All problems seem to them to be due to other people not being as wise or as noble as they are."[35]

To make matters worse, there is actually quite a bias against diversity of thought on most university campuses. For instance, there was a time when "critical thinking" meant giving all ideas— even ideas you are predisposed to disagree with—a fair hearing. Today, when the typical progressive professor urges students to be critical thinkers, what the professor really means is that students should learn effective techniques for criticizing traditional and conservative ideas and values. Thus, "critical thinking" has become newspeak for "correct thinking." The lack of subtlety and the penalty for nonconformance don't escape most students.

Too often, classrooms are ruled by the biases of individual professors. Students suffer when truth becomes a matter of the professor's opinion. Many professors have bought the argument that because perfect objectivity is unattainable all attempts at objectivity are misleading. So we should just get on with spreading our favorite propaganda. As the popular saying goes, "Don't confuse me with the facts, my mind is already made up." Or as the troubled youth Cole whispered in the movie *The Sixth Sense*, "They only see what they want to see."

Educators are often the first to say that schools should teach students *how* to think, not *what* to think. Yet, this simple declaration is often used as an excuse not to teach facts. Whether educators can teach students how to think as a generic process is doubtful. However, when schools engage in indoctrination, the antidote is clear: students should do their own reading and investigating outside the classroom. Students who make a habit of digging deeper into topics will be better informed and more saleable to many if not most potential employers, because businesses and even non-profit organizations are always looking for people with the initiative and experience to solve problems.

Students should be encouraged to explore a wide variety of ideas and opinions. In today's progressive-dominated education that often means reading authors that their teachers have never read or even heard of such as popular conservative economist Thomas Sowell (quoted above), prolific conservative historian David Barton, and veteran climate scientist Roy W. Spencer. Though many professors may not know who these eminently qualified and knowledgeable authors are, businesses seek employees who understand that there are at least two sides to every issue.

If a college does not offer opportunities for participating in the local business community, then students should take it upon themselves to secure valuable work experience—even if that means volunteering to gain the experience. Jobsites offer students opportunities for expanding their horizons.

So as recent graduates make their way beyond the tossed tassel, they should beware: today's world has less esteem for higher education. Prospective employers are looking for graduates whose education includes a balance of liberal arts and technical specialties, real-world work experience, respect for differing perspectives, and wariness toward simplistic or formulaic solutions.

Stilted View from the Academy

A modern perspective on science education and science in general is presented in *Nonsense on Stilts: How to Tell Science from Bunk*[36] by City University of New York philosophy professor Massimo Pigliucci. Unfortunately, it's easier to see through the discarded myths of the past than today's science fads.

The view from the ivory tower has become increasingly obscure and gloomy. By this time, the great unwashed masses should recognize and look up to the leading scholastic luminaries in a range of disciplines. Instead, the hoi polloi have come to rely on an army of non-academic figures to advise them on the important scientific issues of the day.

Nonsense on Stilts is a very approachable book about what's going on in both science and the philosophy of science. The important and complex relationship between these two fields is illuminated in a way that anyone with a passing interest will find interesting. Pigliucci presents in fascinating detail the controversies and personalities behind the advance of scientific thinking and practice over the centuries. For instance, Mr. Pigliucci offers a wonderfully lucid discussion about the difference between deduction and induction in scientific reasoning. More important, he provides a terrific overview of scientific theory and practice.A combination of induction, deduction, intuition, and perspective are found on the positive side of the ledger. A mix of superstition, mythology, postmodernism (the belief that objective knowledge is impossible), and constructivism (the belief that scientists construct knowledge about the world) are found on the negative side. In particular, Pigliucci's discussion about educators' unwarranted love affair with constructivism is refreshing.

Unfortunately, *Nonsense on Stilts*, almost completely ignores the modern practice of science in the world beyond the campus. Perhaps it is ignored because most career academicians lack practical experience and are uneasy with how freely opinions are offered and challenged in the outside world. Naturally, this unease tends to limit perspectives on complex issues—a limitation most clearly visible in Mr. Pigliucci's treatment of global warming.

For all of his talk about how to tell the difference between science and bunk, the author accepts anthropogenic global warming based on his beliefs that the Earth "really does work very much like a greenhouse" and that the 2007 IPCC report must have been moderate because it was "even endorsed by the administration of George W. Bush..." All of which goes to show that just because an author writes about superstition and mythology, that's no guarantee that he won't fall prey to the very same mistakes.

In fact, myths and superstitions are most likely to put down roots where there is little or no resistance. That's why science needs dissenters. Even when they are wrong, they keep everyone else on their toes. When they are right, they may be the only ones keeping the truth alive in an otherwise hostile environment.

Unfortunately, Mr. Pigliucci fails to recognize that the word "pseudoscience" is often used to dismiss anyone who challenges current science establishment orthodoxy from a legitimate scientific perspective. Many scientists in academia ignore the insights of atmospheric scientists practicing outside the academy and dismiss them as pseudoscientists whenever they challenge prevailing academic opinion. In fact, Mr. Pigliucci himself denounces those who challenge the predominantly academic view of anthropogenic climate change as either pseudoscientists or politically motivated. In the process, he dismisses the experience and critical thinking skills of a large number of atmospheric science practitioners who are sincerely unconvinced of humanity's long-term role in global climate metamorphosis. Ironically, this puts Mr. Pigliucci in the position of defending an hypothesis that may itself go down in history as yet another "scientific blunder."

There has to be a clear standard for what constitutes a "scientific blunder." It can't just be a matter of popularity or fashion, because history is flush with scientific blunders that enjoyed widespread—even consensus—support. Today, the only thing we can say with assurance about human-induced climate change is that neither side knows the entire truth about climate change. Until we have definitive proof, we should encourage further research and debate. It's precisely when one side tries to silence the other that scientific blunders are most likely to prevail.

As economist Thomas Sowell stressed in his 2009 book *Intellectuals and Society*[37], "The population at large may have vastly more *total* knowledge—in the mundane sense—than the elites, even if that knowledge is scattered in individually unimpressive fragments among the vast numbers of people" (emphasis in original). Surely this insight applies as well to climate science, where a large number of atmospheric scientists working in the field have serious doubts about a theory that is mainly touted by scientists working in academia.

Though *Nonsense on Stilts* is a good book to read and contemplate, Mr. Pigliucci should heed his own warning. In the

Introduction, he encourages readers to use his work as "a springboard toward even more readings and discussions, to form a habit of always questioning with an open mind and of constantly demanding evidence for whatever assertion is being made by whatever self-professed authority—including of course yours truly."

I suggest this crucial first step to better understanding: The knowledge and insights of *both* researchers working in academia and scientists practicing outside of academia should be taken into account. It's no coincidence that people use the words "ivory tower" when they want to convey aloofness and insularity.

SideBar11: The Academic Echo Chamber

The largest echo chamber on earth is probably the academic echo chamber. On most college campuses, within the vast hallowed halls protected by ivory towers, lies an insular society that shields progressive educators from experiencing the hard facts of the messy outside world. Inside the marble walls, fellow faculty and compliant students will construct ethereal edifices and buttress them with complementary concoctions. The college community offers its own absolutes of *up and down, good and bad, black and white*. And, woe to anyone who dares to dissent.

When Naomi Schaefer Riley, a commentator hired to provide conservative views to the blog connected with *The Chronicle of Higher Education*, dared to question aspects of doctoral dissertations from African-American studies departments, the echo chamber reverberated with a cacophony of hundreds of irate academics. A common, predictable accusation was that Naomi was a racist. Obviously—she was white and dared to criticize something black. She was eventually fired to appease the flustered faculty. Too bad nobody thought to ask Naomi's African-American husband whether he thought she was a racist. Although, since Jason Riley is a member of the editorial board of *The Wall*

Street Journal, maybe his testimony wouldn't count with the self-enlightened elite.

Aligned with predictable group-think, the academy can sell scholastic soothsaying with unabashed certainty. In the 1960s and 70s, it was worldwide disaster from over-population and global cooling that were the pandered prophesies. But, since the news media was a little more objective back then and the public much more level-headed, many of the lofty emanations were tempered by skeptical listeners.

Today, as new claims are made with even more complacence and a compliant media, the bewildered public has a hard time figuring out fact from fiction. For instance, although global temperatures continue to level off, belying the fashionable global warming predicted by vaunted climate models, strident claims of climate catastrophe continue to be inculcated within and beyond the college campus. The mantra seems to be that every time temperatures go up, humans are to blame; and, every time temperatures go down, it's nature offsetting humanity's adverse atmospheric affect. Apparently, changes that occur naturally are good, changes that occur anthropogenically are bad.

Yet, whatever happened to good old commonsense? Like challenging dubious dissertations as Ms. Riley did, has it become anathema to question audacious prognostications such as those that detail global temperature patterns out to the end of this century? Are you unreasonable because you doubt that even really, really smart professors can see distant doom based on their confidence in their own abilities to model the complex climate?

Or, have you simply stepped outside the academic echo chamber long enough to hear rational alternate voices with a bit less distortion?

Minority Ideology Drives Majority Climate Science

WILLIE SOON OF Harvard's Smithsonian Center for Astrophysics, John Christy of the University of Alabama, and many other honest, objective scientists must be doing something right if their legitimate challenges to "settled" climate science are under fierce attack by bellicose liberal politicians. Not only are the Natural Resource Committee's ranking Democrat, Rep. Raul Grijalva, and some like-minded colleagues hounding such heretics, but so are the President and his Environmental Protection Agency. High honor indeed.

Apparently, an ideologically-fueled cohort is driven to wreck some of the most outspoken, qualified challengers to the belief that lowly people are highly responsible for long-term climate catastrophe. Make no mistake, this belief— and considering the prophetic aspects of the anthropogenic climate change claim, it is a belief—is largely propagated by a liberal (that is, left-leaning or progressive) mindset.

Yet, the latest Gallup poll concluded that conservatives "continued to outnumber moderates and liberals in the U.S. population in 2014, as they have since 2009." In fact, from the early 90s, when Gallup polling on this self-identified political ideology comparison began, the percentage of conservatives and moderates was typically about double that of liberals. And, although the liberal percentage has been creeping up over the years to its highest on record at 24 percent last year, conservatives and moderates combined still command a sizable majority of the ideology of the U.S. population.

Consider that the control of ideas, especially in key areas such as journalism, entertainment, education, and even to a substantial degree, science, is solidly in the liberal camp.

Take education and the field of environmental science for example.

From grade school through graduate school, practically all students are subjected to purely liberal bias on environmental topics. Whether the subject realm is in the biosphere, atmosphere, hydrosphere, or lithosphere, the challenges to the natural world and their solutions are confined within the liberal framework. Free-market environmentalism, wise-use movement, or other conservative or even libertarian approaches are regularly distained or completely ignored.

It's not surprising then that educated people graduate with an inculcated liberal bias relative to environmental issues. This bias is reinforced beyond the campus by most media reporters, who are themselves disciples of the education establishment.

The field perhaps most regrettably saturated with liberal groupthink is the field of science itself. This situation is bad enough alone, but with the support of liberal journalism and academia, the problem is expanded by an order of magnitude.

Foregone conclusions, such as *humans are largely responsible for long-term, disastrous, global climate change*, are readily accepted. The disincentive to disagree with this conclusion is terrific. And, with incentive from the White House's climate change action plan, the Environmental Protection Agency's Clean Power Plan, and truckloads of federal funding, it's hard not to become part of the campus carpool.

In addition, there is a strong incentive from peer pressure.

The March 2015 issue of *National Geographic* magazine unintentionally states it well with their cover story on "The War on Science." In the related article by *Washington Post* science writer Joel Achenbach, Joel states that science "appeals to our rational brain, but our beliefs are motivated largely by emotion, and the biggest motivation is remaining tight with our peers." He goes on to quote the current editor of *Science* journal, geophysicist Marcia McNutt, as saying, "We're all in high school. We've never left high school....People still have a need to fit in, and that need to fit in is so strong that local values and local opinions are always trumping science. And they will continue to trump science, especially when there is no clear downside to ignoring science." Marcia certainly did not intend this quote to apply to scientists themselves, but why not?

Academic scientists, especially, are in an echo chamber (see sidebar in previous chapter); and, if they went from high school to undergraduate then to graduate school, who's to say they "never left high school" with respect to group pressure and thinking?

So, for instance, when actual climate data for more than a decade and a half belies government and academy predictions of global warming, unbiased scientists would start looking around for a new or substantially revised hypothesis on climate change. Scientists saturated with a liberal viewpoint would continue to ardently defend a patently disproven one.

As it becomes more apparent that long-term global climate is resilient to human activity, hopefully conservative influence on the highway of ideas will become fast tracked. If not, at least the conservative and moderate majority should be driving more of the discussion. And, aggressive liberals, including those at the highest levels of government, need to more seriously consider the majority perspectives that just may help solve some of the most unyielding environmental and societal challenges now and down the road.

Leftist Science in the Climate-Science Neighborhood

LEFTISTS OR RADICAL socialists (often disguised with the euphemisms "liberals" or "progressives") have many potential avenues they can travel to arrive at their ends—an ends that justifies its means. Typically, the leftist trail is carved out through the down-trodden by whipping up resentment and jealousy and a sense of entitlement that is to be fulfilled by the social activist/community organizer. Furthermore, leftist socialites in high political and organizational authority are celebrated and idealized by the witting and unwitting media as saviors of the needy.

Thus, the attack on society comes from below and above facilitated by the media to ultimately enslave and impoverish all—all except the ruling class of leftists, and the wealthy who carefully avoid excessive irritation of the rulers.

After decades of inroads into unions, education, politics, law, journalism, public relations, and many of the "soft" scientific disciplines, the latest avenue of attack has been via the "hard" sciences, in particular, atmospheric science.

The leftist techniques that have become apparent during my decades of practice in the atmospheric science profession include blatant dishonesty stemming from arrogance—a hallmark of socialism—and its offshoot, a sense of supreme superiority.

Fundamentally, leftist practice has no problem with twisting the truth or simply lying. Although for most people, "honesty is the best policy," in leftist ideology the ends justifies the means and so, even though verity can be useful, it's not absolutely necessary

For instance, there are many experienced real-world atmospheric-science practitioners who legitimately question the conclusions of the cadre of academic and government scientists who have declared "settled" the complex scientific endeavor of understanding climate change. When experienced practitioners are labeled "climate change deniers" by some of these very same arrogant scientists, you know leftist ideology has reared its ugly head.

The denier moniker is obviously untrue, which makes it problematic to trust a researcher's or research director's work who relies on using this blatant *ad hominem* attack. As disclosed in another chapter, professor Michael Mann used the phrase "climate change denier" or some variant of it seven times on one page alone, page 193 (if you count endnotes to the page), in his 2012 book *The Hockey Stick and the Climate Wars: Dispatches from the Front Lines.* Secretary of the Interior, Sally Jewell, impudently stated in August 2013 that she didn't want any "climate-change deniers in my department."

Furthermore, another basic tactic of leftist practice is to *divide and conquer*, that's why differing science views are framed as a "war." It's all or nothing, whereas with authentic scientific practice a middle ground is quite acceptable and often is the path to discovering the truth about a matter under investigation. Of course, with extremists on both sides, "you're either for us or against us," a view more typical of politics than science. But, leftists demand that you choose sides—pick a conclusion and defend it at all costs. With such leftists in control of academia and much of the media, a scientist's conclusions must meet leftist approval or he/she is not "doing science." Independent thinking is not acceptable. And, if you have doubts that the leftist side is correct, then you must be a denier of the "truth."

Since the leftist ideology is all about its own power, it doesn't take too kindly to independence and independent thinking in others.

Much of the higher-education mindset is steeped in progressivism. Within the academic hallowed halls—perhaps the largest echo chamber on earth—leftist groupthink trumps independent thinking, no matter how much that independent thinking is reasonably based on application of theory in wide practical experience and objective commonsense.

Unfortunately, the concept of man-made global warming emerged from the college campus in the form of high-brow theory and is therefore not to be challenged. Challenge is met with traditional leftist tactics, such as *ad hominem* attacks in the form of name-calling and straw-man arguments (like "climate-change denier"), generation of enemies lists (see for example the "Research Database" at the PR spinternet site *DeSmogBlog.com*), and even "storm" troopers (like the National Center for Science Education that used to exclusively hunt down those who dared to question molecules-to-man evolution but has now expanded its crusade to climate heretics).

Climate-change challengers also face exclusion and missed opportunities (since any chance at securing grant money from a government that dictates what is worthy of research is through acceptance of the prevailing "humans-did-it" hypothesis), and squelched career advancement (if you have *any* doubts that people are causing a climate catastrophe but want to stand-out in the climate field, stay in the closet!).

Scientific practice is a kind of social construct with both technical and philosophical aspects; you don't have science in practice without a philosophy of science. Because philosophy today is generally developed within the college community, the current philosophy of science is based in the progressivism running amok through much of the collegiate commune. Thus, such philosophy applied to the understanding of climate change requires that you conform to the idea that human activity is regulating the long-term global temperature.

Now, every climate-change statistic must be filtered through the leftist lens. If global temperatures increase (or decrease) by a smidgen this year, humans are to blame. If ice melts at the North Pole (or expands at the South Pole), curse your internal combustion engine. If we don't get enough (or get too much) snowfall during the winter, scowl at your neighbor's fireplace. If a drought persists in the Midwest (or the Mississippi River floods), people are fiddling with the weather dials. If tornado or hurricane frequency or intensity increases (or decreases), batten down the hatches with scorn for fossil-fuel users.

With the imposition of leftist ideology on climate-science hypothesizing, foregone conclusions are inevitable; and, woe to anyone who challenges those conclusions. Such predetermination produces the leftist classic results of impoverishment and enslavement, which can be seen in the form of higher energy costs and limited economic choices for all.

So, although the academic leftists' actions are rather juvenile and result from arrogance, self-righteousness, exclusivity, and narrow-mindedness, they are to be taken seriously.

The negative consequences of leftist thinking can be particularly severe, disseminated by those who presently wield power on campus and beyond to the highest levels of government. But, the freedom of independent thinking is one of the hallmarks and joys of authentic science. This thinking is what helps science to make great advancements and helps society to truly progress.

Above all, science should be in the service of humanity, rather than in the service of any ideology. But, unfortunately, with leftist practice now taking up residence on Atmospheric Science Street, well... there goes the neighborhood.

SideBar13: Eliminating the Left's Euphemisms

Here's good advice for conservatives, as this political campaign season is revving up: Don't use the euphemisms "liberal" or "progressive" to describe those on the left. These descriptors immediately cede a perception advantage to your challenger. The better, more accurate term is "leftist," since those on the political left are typically not liberal in the true sense of the word, nor are they progressive.

Good examples of this come from the issue involving climate science. Leftists have worked long and hard to shutdown reasonable debate on the question of the degree of human influence on climate change. Thus, leftists are certainly not being liberal regarding exchange of ideas.

In addition, leftist solutions to energy needs of nations is to fight supposed anthropogenic global warming with yesteryear's sunbeam and sea-breeze catchers, rather than to get on board with modern, cutting-edge fracking and nuclear technologies. Thus, leftists are not being progressive regarding advancement of society.

Unfortunately, leftist cannot even be counted on to liberate and develop people by championing true freedom, since leftists look to impose their views on others directly or through government control. So, if conservatives are looking for an alternative moniker for leftists, perhaps "statists" fits the bill.

Climate Change in Perspective

Every year, with the fanfare of released official weather records comes the usual claims of "warmest this" and "historic extreme that." But, facts are stubborn things.

Global temperatures have leveled off over the past 18 years, even as worldwide carbon dioxide (CO_2) concentrations have continued to rise. This is a measured fact that confounds strident climate-change predictions. Why no corresponding rise of temperatures with increasing CO_2? Because other gases, vapors, and chemical/physical effects are still operating in the atmosphere to influence climate. CO_2 does not act in a vacuum.

Thus, as CO_2 has increased dramatically over the previous decades, substantial global temperature increases have not materialized as we were assured they would by academic, government, and political scientists. This is in large part because of the overwhelming role water plays in regulating climate. For example, while CO_2 is only at 0.04% by volume in the atmosphere, water vapor ranges from about 0 to 4%. Water, in the vapor phase and in the liquid and solid phases (besides the energy associated with the phase changes of water, and oceanic circulations), is the biggest climate controller on earth.

As far as climate statistics go, much perspective is needed. First of all, with respect to just about every recent year being touted as the warmest year on record for the contiguous U.S., note that the contiguous U.S. covers less than 2% of the earth's surface.

Next, the increase in globally-averaged temperature of about 1°F is quite small by comparison with the mid-20th century baseline. I was an official weather observer early in my atmospheric-science career (starting more than 35 years ago). I know from personal experience that temperature measurements, which were typically made once per hour, were made by "eyeball averages" based on where the top of the liquid stood in the thermometer's glass tube. So, in practice, measurements were made within about + or - 0.5°F accuracy. Notably, a 1°F increase is small indeed, taking measurement limitations into account.

Furthermore, the fact that the recent levels of the average global temperature have been a little bit higher than the average since reliable land-surface records have been continuously kept (since the mid- to late- 1800s) is also not remarkable. When good consistent records began, the globe was emerging from the latest of multiple ice ages. From approximately 1550 to 1850 the globe was experiencing what climatologist traditionally called the "Little Ice Age."

And, the claim that any particular recent year was a historic year for extreme weather... well, that is pure opinion, based on convenient myopia. Expanding the field of view, we see that, in the U.S., there were the Dust Bowl years in the 1930s, extended drought from 1949 to 1956, and massive historic New England storms in 1938, 1893, and 1821, to name a few memorable incidents. In addition, we can't be blind to the fact that some recent years, like 2012, had a near record low amount of tornadoes.

As for melting ice in the far north, while a substantial arctic storm contributed to the record minimum in 2012 Arctic sea ice extent by blowing apart much of the ice, in the southern hemisphere, the Antarctic ice sheet has grown in recent years.

Climate anomalies, which can be selected to "prove" rising or falling thermometers, happen all over the globe. Picking changes that just match the current academic, governmental, and political "consensus" on climate change is not how authentic science is supposed to work.

But, it's no surprise that the consensus of scientists who are paid huge sums of money from government find that humans are responsible for climate change. After all, as the U.N.'s IPCC puts it, the job today in the world of climate science is to understand the scientific basis of "the risk of human-induced climate change, its potential impacts and options for adaptation and mitigation." Thousands of the rest of us atmospheric scientists, who have no stake in the outcome, find the hypothesis that humans are responsible for long-term, catastrophic, global climate change to be a stretch at best.

Perspective on the risk and benefit of CO_2 emissions effect on the atmosphere is desperately needed. Regardless of the theoretical risk that CO_2 from fossil-fuel burning for energy production is the driver of long-term, dangerous, global climate change, the benefits of the use of such inexpensive fuel to alleviating miserable living conditions for about one billion people are clear. Yet, a trillion dollars may be spent over the next decade to reduce, eliminate, or sequester CO_2 emissions that may yield only a fraction of a degree Celsius decrease in global temperatures (if you believe the climate-prediction models). Sadly, such a spending spree will do next to nothing to address real tragedies like abject poverty in a world of plenty.

So, regardless of the heavy handedness from the Obama administration on supposed manmade global warming, thousands of knowledgeable scientists are not convinced of human culpability for long-term dire climate change.

Many of us have witnessed the increasing distortion of the science behind climate change and especially the arrogance behind long-range global warming predictions.

Over several decades now, the climate around the world has

changed a lot. It's gone from cold to hot to now back to cold again (*maybe*). But, I'm not talking about the meteorological conditions. Rather, I'm referring to the climate of fear promoted by leftist ideologues, such as those currently in charge at the highest levels of government, who use the vagaries of the weather to promote their left-leaning agenda.

Early on, in the mid 1970s, the talk and teaching was about an impending ice age. Prompted by downward global temperature trends from the recent past and popular academic theory, the expectation was that the earth was headed for some icy climes. The "Milankovich Theory" was popular at the time—a theory proposed by astrophysicist Milutin Milankovitch. He posited that Earth's climate was strongly influenced by cyclical variations of Earth's orbit around the sun and the planet was in for some chilly changes.

In the 1980s, as the mercury started to rise worldwide, so did the rhetoric for a supercharged "greenhouse effect," fueled by too much greenhouse gas. The blather came to a head when planetary scientist James Hansen, Director of NASA's Goddard Institute for Space Studies, testified before a special congressional hearing held by Democratic Senator Timothy Wirth on June 23, 1988. Dr. Hansen accused human activity of disproportionately heating the atmosphere. And, as noted previously, with the formation of the U.N.'s IPCC that year, hothouse hysterics was off and running.

Fast forward to today, with an 18-year lull in any substantial global temperature rise, there's some talk that the mercury may be headed downward again (see for instance, "A Paradigm Change: Re-directing public concern from Global Warming to Global Cooling, by veteran climate scientist S. Fred Singer in *American Thinker* on July 31, 2015). Will the hype now turn back to pushing a new ice age brought on by too many people, too much energy use, or too many Republicans?

The thermal hiatus prompted Matt Ridley, the prolific science writer and member of the British House of Lords, to pen an op-ed in *The Wall Street Journal* (September 5, 2014) that asked "Whatever happened to global warming?" The answer to this question is being investigated by the "settled" scientific community. However, the answer to a related question, "Whatever happened to climate science?," can be readily answered.

The pure science in climate science went missing when politics became a factor in the forecasting equations. The political atmosphere originated largely from organizations like the U.N.'s IPCC and their weighting of man-made carbon dioxide as a key ingredient in long-term global climate change.

Politics continues to distort scientific practice by requiring lopsided analysis on, for example, the Administration's inherently biased "Social Cost of Carbon." Whatever happened to the social *benefit* of carbon? After all, for a proper perspective, the science of economics considers cost-benefit analysis to be standard practice when determining the value of a project or prospect. For balance, we can point to empirical data that has confirmed that since plants thrive on carbon dioxide one obvious benefit of that increased gas in the atmosphere has been a greener globe.

Nevertheless, the U.S. Environmental Protection Agency (EPA) also throws in with twisting scientific conclusions. Consider that in 2014, the EPA announced in the *Federal Register* that it is looking for research participants for their new fight against the public-health impacts of climate change (ostensibly human-caused). Healthy skeptics need not apply.

Optimism for sunnier conditions will come when science is not politicized, when scientists practice humility, and when they are free to explore any hypothesis, theory, or doubt that moves them.

SideBar14: Informed Perspective

In my office, I have posted three opinion pieces from *The Wall Street Journal* that intelligently address some of the oft repeated, superficial claims of those championing the anthropogenic climate change hypothesis. The *Wall Street Journal* commentaries are:

"The Myth of the Climate Change '97%'," May 27, 2014, by Joseph Bast, president of Heartland Institute and Roy Spencer principal research scientist for the University of Alabama in Huntsville;

"Whatever Happened to Global Warming?," September 5, 2014, by Matt Ridley, veteran science writer; and

"The Alarming Thing About Climate Alarmism," February 2, 2015, by Bjorn Lomborg, director of the Copenhagen Consensus Center.

These opinion pieces provide much needed perspective on the climate change issue.

In addition, John Steele Gordon penned "The Unsettling, Anti-Science Certitude on Global Warming" in *The Wall Street Journal* on July 31, 2015. The op-ed succinctly addresses two major bits of oft-spouted nonsense in the climate debate arena. Gordon's opinion begins with:

"Are there any phrases in today's political lexicon more obnoxious than 'the science is settled' and 'climate-change' deniers?

The first is an oxymoron. By definition, science is never settled. It is always subject to change in the light of new evidence. The second phrase is nothing but an ad hominem attack, meant to evoke 'Holocaust deniers'...

The phrases are in no way applicable to the science of Earth's climate. The climate is an enormously complex system, with a very large number of inputs and outputs, many of which we don't fully understand—and some we may well not even know about yet. To note this, and to observe that there is much contradictory evidence of assertions of a coming global-warming catastrophe, isn't to 'deny' anything; it is to state a fact. In other words, the science is unsettled—to say that we have it all wrapped up is itself a form of denial. The essence of scientific inquiry is the assumption that there is always more to learn."

Anti-science nonsense can lead to sad consequences. In the August 13, 2015 lead editorial, "Obama's Climate Plan and Poverty," the editors of *The Wall Street Journal* state the anticipated effects of the administration's Clean Power Plan:

"As energy prices rise [as a result of the Plan], they spill into other basic needs like food, via fertilizer and feed, and housing, via building materials like cement. Everyone ends up with less disposable income and a diminished standard of living, but low-income workers really are worst off."

Beyond informed opinion, every year now since 2008, the Nongovernmental International Panel on Climate Change (NIPCC) has been a worthy challenger to the United Nation's Intergovernmental Panel on Climate Change (IPCC). The NIPCC is "an international panel of *nongovernment* scientists and scholars who have come together to understand the causes and consequences of climate change" (quote from NIPCC's website, emphasis mine). The level-headed scientific documents produced by this organization and its conferences provide a point of view sorely needed in today's contentious atmosphere. Scientific reports associated with the NIPCC include:

Climate Change Reconsidered (June 2009)
Climate Change Reconsidered II: Physical Science (September 2013)
Climate Change Reconsidered II: Biological Impacts (March 2014)

Each of these reports, containing articles by well-qualified researchers, is a direct challenge to the U.N.'s IPCC declared conclusion which is to understand the scientific basis of "the risk of *human-induced climate change*, its potential impacts and options for adaptation and mitigation" (emphasis mine).

Slapshot Science

IN HOCKEY, "[T]HE slapshot is harder than other shots, and because of the violent motion involved, somewhat less accurate. It also takes longer to execute; a player usually cannot take a slapshot while under any significant pressure from an opposing player..." [38]

Colleges across America have abandoned pure science in favor of ideological science. And it's all because of a hockey stick.

Perhaps no climate science icon is more recognizable than the "hockey stick" curve originally produced by Dr. Michael E. Mann and his colleagues in the 1990s and published in the prestigious journal *Nature* in 1998. This graph became the featured diagram in the 2001 IPCC *Third Assessment Report*. (See simplified renderings of this graph compared with the traditional historic temperature graph for 1000 to 2000 A.D. in Figure 1.) The graph was instrumental in convincing many government policymakers that human-related emissions of carbon dioxide are causing an unprecedented increase in global temperatures and that drastic action is once again needed to save the planet.

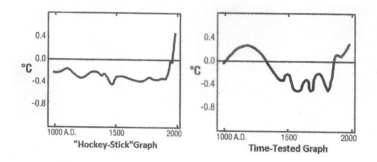

"Hockey-Stick"Graph Time-Tested Graph

The "hockey stick" graph (left) denies the dramatic medieval warm period and little ice age that were manifest in the long accepted historic temperature graph (right)

In *The Hockey Stick and the Climate Wars: Dispatches from the Front Lines*,[39] Dr. Michael Mann, professor and director of the Earth System Science Center at Penn State University, proffers a spirited defense of his graph reconstructing temperatures over the past 1000 years. Dr. Mann repeatedly stresses that there are numerous representations of proxy data that produce the same tell-tale pattern. He ably demonstrates that challenges to his work, whether based on his selection of data or statistical analysis, are feeble at best. And he personally relates in painful detail how political attacks are harmful to free inquiry. As a general defense of academic scientific research into recent climate change, *The Hockey Stick and the Climate Wars* is a wonderful book, brimming with peer reviewed references and reasoned arguments. As a hockey sportscaster might exclaim, "*He shoots...he scores!*"

Dr. Mann's book will undoubtedly be quoted and referenced for years to come in support of the academic science establishment. But therein lies the book's major weakness: it represents a narrow and insular perspective. Michael Mann apparently believes that his interpretation is so obviously correct that challengers could not possibly be acting in good faith. He resorts to unprofessional name calling—relishing the accusatory phrase "climate change denier"—throughout *The Hockey Stick and the Climate Wars*. On one page alone (page 193, including the associated end notes), Mann uses this ad hominem attack or some variant of it seven times. The term is patently absurd because no one denies that the Earth's climate changes. And it's also disgraceful: Dr. Mann and others use this expression in an attempt to equate recalcitrant scientists to holocaust deniers—as if challenging anthropogenic global warming dogma is tantamount to covering up one of the most heinous crimes in history.

Unfortunately, Dr. Mann did not draw the proper conclusion when others tried to intimidate him into silence. Instead of recognizing that scientific progress requires the freedom to challenge and debate ideas, Dr. Mann apparently concluded that it is better to silence others than to be silenced. If his ideas are correct, then he should welcome challenges; ideas that withstand repeated criticism gain tremendous respect. By likening his critics to Holocaust deniers, Dr. Mann demonstrates his insecurity and belief in free inquiry only for himself and like-minded professors. The problem, however, is not just that Dr. Mann fails to treat others as he would like others to treat him. Rather, by calling skeptics "climate change deniers" he smears those who dare to question science establishment orthodoxy. He's not taking a courageous personal stand—he's simply participating in "everybody does it" behavior.

Dr. Mann relies heavily on progressive (i.e., politically-motivated) sources such as Media Transparency (the part of the Media Matters for America action network "dedicated to analyzing and correcting conservative misinformation"), Government Accountability Project, Sourcewatch.org, DeSmogBlog.com, Union of Concerned Scientists, *American Prospect*, Center for American Progress, Think Progress, ad nauseam. What these groups have in common is that they believe that science may be politicized—that it's okay to spin science for their causes.

The book's arrogant tone, so typical of progressive academia, provides a hint as to why the public is increasingly suspicious of institutions such as Congress, the media, and the education system. And perhaps that's why today's *ex cathedra* climate science proclamations are encountering so much resistance from mere mortals. Rushing to the rescue, Dr. Mann announces the creation of the Climate Science Rapid Response Team (essentially a "truth squad")[40] to quickly quash any threat to science establishment ideology from, for example, those troublesome atmospheric scientists who ply their trade in the real world.

Even if Dr. Mann's reconstruction of climate conditions over the last thousand years is an accurate representation of climate change (which, based on the limited proxy data, is arguable), the climate science authorities' insistence that human-produced emissions will cause severe meteorological mischief across the globe through the end of this century and beyond is pure prognostic presumption. The Earth's climate system is highly complex, and pretending that it is controlled by a few simple knobs and levers doesn't make it so. Even the role of the principal climate regulator, water—manifested as a solid in ice sheets, a liquid in cloud droplets and oceans, and a vapor in ambient air—is not sufficiently understood for scientists armed with climate models to become trusted fortune-tellers.

While it's not clear that current human activity can substantially change the Earth's climate, it is clear that current human activity is changing the field of climatology.

Earlier, I lamented the intrusion of politics and ideology into the scientific debate concerning climate change. Now, politics and ideology have found their way into the newest editions of university science textbooks. It would be nice to think that students exposed to this revisionist science will, with the help of knowledgeable professors and a little independent thinking, be able to see through the deception. Sadly, many in academia treat the ideas presented in textbooks as the gospel truth, and there is a real danger that generations of students will be indoctrinated by activists' conclusions about scientific issues, rather than educated by being presented with the messy facts and the chance to sort them out on their own.

Here's a specific example from a popular climatology textbook.[41] It involves the infamous "hockey stick" curve discussed earlier that purportedly tracks global temperature fluctuations over the past 1,000 years, with its 900-year long handle leading to a 100-year long angled head of the stick.

The hockey stick graph contradicts the historical temperature graphs that climatology professionals have relied on for decades. In the early days, it was anyone's guess how global temperatures fluctuated in the distant past. In an effort to deduce average annual global temperatures going back 1,000 years or more, some of the field's pioneers tried comparing recent temperature measurements with "proxy" data such as tree rings, polar ice layers, lake bed sediment strata, and written records. Data gleaned by numerous atmospheric scientists and published in hundreds of scientific papers revealed a distinct medieval warm period from roughly 950 to 1250 A.D., which was dubbed the Little Climatic Optimum. Scientists also documented a significant cooling period between 1645 and 1715 that corresponded to an apparent period of sunspot inactivity and was called the Maunder Minimum (after Edward W. Maunder, the British astronomer who first studied this period). The climatology community believed that the two periods of relatively extreme temperatures had actually occurred, and climatology textbooks included historical temperature charts displaying the two periods.

These dramatic swings, representing periods of warming and cooling on the original charts, somehow disappeared with the introduction of the hockey stick graph. How could this be? An investigation by researchers Stephen McIntyre and Ross McKitrick looked at how the raw proxy data was selected and subjected to statistical analysis to create the hockey stick curve.[42] Unfortunately, for McIntyre and McKitrick, Dr. Mann makes a good case that the investigation was probably flawed.[43] Regardless, by 2005 the hockey stick graph was considered equivocal by many scientists, but that didn't stop the IPCC from continuing to use it in some form along with complementary graphs in subsequent reports.

Despite the controversy surrounding the revision of past climate conditions, the editorial team tasked with producing the 2010 edition of the textbook mentioned above performed a nifty switcheroo. In the new edition, the hockey stick graph replaced the original temperature graph without a word of explanation. Recall that the original graph was the field's accepted and time-tested illustration of temperature changes over the past one-thousand years. Certainly the sudden replacement of the original graph with a new graph having profound implications merited discussion. In fact, a side-by-side comparison of the two graphs along with an explanation for preferring the new graph would have made for good science education.

There are legitimate reasons for discarding a scientific theory, but discarding a theory because it conflicts with a particular political agenda isn't one of them. Nor is it appropriate for educators to present contested data and controversial theories to students as consensus views without acknowledging and fairly describing the main dissenting views. As Daniel Patrick Moynihan said, "Everyone is entitled to his own opinion, but not his own facts." Science, more than any other discipline, should maintain a bright red line between undisputed facts and everything else—including all theories, no matter how popular they may be.

The scientists promoting the hockey stick graph with its

revisionist climate history may sincerely believe that they are doing the right thing. However, it's one thing to promote your ideas and another to erase other ideas and treat further discussion as out-of-bounds. That's not the way science is supposed to be done. This case is particularly damaging to climatology, but it's also damaging to science in general.

In fact, the damage may spill out beyond science. Science as taught in the classroom eventually makes its way into the community. Citizens or their elected representatives are frequently asked to consider and even vote on proposals that involve scientific matters. Whether it's about restricting carbon dioxide emissions or installing a new natural gas well, citizens should be given a fair chance at evaluating the data and differing views.

Alas, if temperature records are reworked with an impressive-looking slapshot and voters are wowed by a particular goal, then our planet is indeed in trouble.

The Madness of Environmentalism
Exposed

A COUPLE OF books came out in 2015 that presented a good case for a more conservative perspective on climate change. This perspective is sorely needed right now because the U.S. EPA is apparently operating under the command and control of President Obama's leftist ideology (see related chapter). There is little doubt about this as the president's hand-picked EPA administrator, Gina McCarthy, has basically admitted to professional audiences that she is doing the bidding of her boss. What may surprise folks is that this is not inappropriate with respect to how the EPA was initially established by President Richard Nixon; although, relative to the advancement of the country's economic, environmental, and public health, and the wellbeing of objective scientific practice itself, an ideology-driven EPA is quite inappropriate.

Dr. Alan Carlin, in his impressive work Environmentalism Gone Mad: How a Sierra Club Activist and Senior EPA Analyst Discovered a Radical Green Energy Fantasy (Stairway Press, 2015), explains that the EPA:

> ...reports directly to the President and thus has no independence from the Executive Branch like some regulatory agencies. This means that if an Administration

> *wants to use its power to determine regulations, it can impose exactly what it wishes to do subject only to the Congressional Review Act and Congress' powers of appropriations, both of which have proved ineffective so far in preventing Obama from doing what he wants with regard to EPA.*

Carlin was at the EPA almost from its inception in 1970. He came from research work for the RAND Corporation in Santa Monica, CA, to work with the EPA in Washington, D.C. from 1971 to 2010.

In early 2009, after submitting serious negative comments on EPA's draft Technical Support Document for the Endangerment Finding relative to the adverse effects of increasing levels of atmospheric greenhouse gases, Carlin had been maligned by the EPA powers-that-be for challenging the Obama administration's poor economics and science represented in these findings. Yet, as an EPA senior analyst with an undergraduate degree in physics from Caltech bolstered by a Ph.D. in economics from MIT, Carlin surely knows his stuff.

Carlin asserts that even if EPA's current effort to control carbon dioxide emissions are successful:

> *...it will not change the climate or extreme weather in any measureable way even though Obama has proclaimed it will. It will simply increase the rates paid for less reliable energy, with lower income Americans bearing most of the burden along with the slow recovery of the US economy.*

Throughout his lengthy personal recounting in *Environmentalism Gone Mad* of the rise and fall of EPA adherence to science over politics, Carlin engages the reader with essential details. These details include not only an insider's perspective on the operation of the EPA but also numerous, specific, and sensible short-term and long-term recommendations on how to "get out of this mess"— a mess largely brought about by the current administration's adherence to radical leftist environmentalism. The need to consider reasonable costs verses benefits in air quality rules (as exemplified in the recent U. S. Supreme Court decision in "Michigan, et al. v. EPA, et al.") is a move encouraged by Carlin.

Good economics and science require a broad perspective, yet when politics (and its concomitant financial control) dominates the mix of viewpoints, the climate changes and usually in an ominous way. As Carlin expressed it in one of his long-term reform recommendations to reduce incentives for EPA managers to follow the administration:

> Besides the normal bureaucratic controls, the pay of all EPA executives and senior analysts are directly determined by Congress and the President. This is unlikely to lead to independent action or thought by these crucial civil service employees. Yet independent analysis is desperately needed if EPA is to reflect good science and economics rather than science determined by their political masters.

Without a doubt, *Environmentalism Gone Mad* is an important book that provides well-informed personal insight into the convoluted world of calamitous climate science promoted by what Carlin calls the "climate-industrial complex" or "CIC." The CIC includes the science elites, mainstream media, environmental groups, leftist politicians and bureaucratic administrators, "green" energy and fuel producers and promoters, PR myth makers (like those labeling knowledgeable skeptics as "deniers"), and others who profit financially, professionally, and personally from foisting a future climate fantasy on a unwary public.

As Carlin observes:

If governments simply stayed out of energy decisions not involving government-owned resources, urgent national security objectives, or actual proven pollution problems and let the markets decide how to meet energy needs, everyone except the CIC would be much better off, including the environment.

Ratepayers and all taxpayers would do well to educate themselves on the inefficient, sometimes unscrupulous, and perhaps most-times counterproductive actions of those obstructing the goal of good, clean, and affordable domestic energy. *Environmentalism Gone Mad* is a good first step in this essential education.

Another recent book that would enlighten serious readers on the climate-change issue is by University of Houston professor Larry Bell, a prolific *Forbes* and *Newsmax* columnist. Dr. Bell's new book, *Scared Witless: Prophets and Profits of Climate Doom* (Stairway Press, 2015) is a sort of follow-up to his previous book on the topic of climate change, *Climate of Corruption: Politics and Power Behind the Global Warming Hoax* (Greenleaf Book Group, 2011).

Much of my positive review of Bell's earlier book can be applied to this new one. In particular:

As massive natural disasters and momentous geopolitical turmoil continue to erupt, the American public [as indicated by a recent Gallup poll] is continuing to put global issues into perspective. 'Climate of Corruption' [and now Scared Witless] is a book that will buttress intelligent decisions about where U.S. dollars and traditional American ingenuity and compassion will do the most good to alleviate the real environmental and man-made calamities that affect the world community.[vi]

[vi] http://www.washingtontimes.com/news/2011/mar/29/redressing-

As the old saying goes, being "forewarned is forearmed." And, like *Climate of Corruption*, the book *Scared Witless* delivers much forewarning on the nature of the climate-hysteria business, especially for those just entering the climate debate market of ideas.

Written in a quite readable, folksy style, *Scared Witless* provides an understanding and perspective much needed for the layperson and professional alike regarding the level of human culpability to climate change. Not that humans have *no* impact on climate, the real question is how much? And then, is human impact enough to overwhelm nature itself?

Five book sections cover the gamut of important topics in the climate-change arena. Sections include: "The Climate Alarm Industry," "Disastrous Speculations," "Colorblind 'Green Energy' Madness," "Regulation Run Amok," and "Truly Scary UN Agendas." In general, the thrust of these sections is to stress the idea that increased carbon dioxide concentrations in the atmosphere will produce both "winners and losers," whether you're talking about increased ocean acidity, higher atmospheric quantities, or just comfort levels of Planet Earth's inhabitants.

Scared Witless tackles the major hysteric speculations proffered by climate alarmists with a thoughtful, measured assessment of climate change. Speculations such as shrinking polar ice caps, drowning polar bears, increasing malaria, and acidification of oceans are countered effectively with references to subject-matter experts and refereed journal articles.

One particular subsection, "For Better and Worse, Climate Changes with No Help From Us," delivers the long view on the global warming issue. Regarding the errant environmentalist "good old days" nostalgia, Bell writes:

---

global-climate-hysteria

Realities going back a few hundred years and more reveal a different picture; one displaying widespread poverty, starvation, disease and hardship. Yes, throughout human history, people have had to adapt to climate changes—some long, some severe, and often unpredictable. They have blamed themselves for bad seasons, believing they had invoked the displeasure of the gods through a variety of offenses. High priests of doom told them so, extracting oaths of fealty and offerings of penance for promised interventions on their behalf. In this regard, at least for some, it seems little has changed. That penance today comes at a very high cost...our present and future national economy.

Bell goes on to minister succinct scientific absolution for the religious guilt trip:

Climate changes and shorter term weather events are the way nature balances itself, move heat and moisture around, and provide motivations for species to evolve.

Once again, professor Bell has provided a valuable tool that can be used in the deconstruction of the superficial edifice of "settled" climate science. *Scared Witless* has put the *prophets and profits of climate doom* on notice—knowledgeable challengers will not go quietly.

SideBar16: User's Guide to Useless (and Harmful) Environmentalism

Libertarian conservative James Delingpole's light-hearted compendium on a serious topic, *The Little Green Book of Eco-Fascism: The Left's Plan to Frighten Your Kids, Drive Up Energy Costs, and Hike*

Your Taxes! (Regnery, 2013), is presented in a concise encyclopedic style with entries from A to Z.

Mr. Delingpole of the *Daily Telegraph*, *The Spectator*, et al., covers the gamut of subjects related to climate-change hype. His humorous handy-dandy tome will have a "Denier" smiling, but will take an "ardent environmentalist who also happens to be an extreme masochist... to heights of ecstasy... never dreamt possible."

Particularly cogent is his definition of "Weather." The difference between "Weather" and "Climate" is really very simple, he explains. "If it's cold when the greenies say it should be warming, it's weather. If, on the other hand, it's doing anything that bears any correlation with their computer models' predictions of environmental doom, then it definitely counts as climate."

Snippets from some of my other favorite topics include:

"Blogosphere": "Not since Gutenberg invented the printing press, perhaps, has the cause of openness and liberty made so spectacular a leap as it has as a result of the internet."

"Ecology": "Ecology, you could argue, has replaced Christianity as the West's dominant religious philosophy." And, the related offering, "Faith": "Modern environmentalism is only comprehensible in terms of faith. It is a form of antinomianism— that is, it depends for existence on a series of accepted 'truths' which brook no argument and are entirely immune to logic or rational criticism."

"Lifestyle, Too Selfish to Change Your": "Charge routinely leveled by greenies at all those reactionary conservatives who: *Insist on taking showers more than once a fortnight...*"

"Scientists": After giving some much-needed perspective on scientists, Mr. Delingpole tackles "science" observing that political activists discovered that science could be used "as a handy excuse to advance their agenda under the guise of studied objectivity.

'Hey, it's not because we're a bunch of crypto-Marxist control freaks that we're demanding higher taxes, more regulation, and the replacement of Western industrial civilization with a Soviet-style global command economy run by leftist technocrats. It's because the science tells us that that's what we need to do.'"

Additional notable entries include "Golden Rice" (not one of the third-world souls suffering from lack of GM-foods "has ever once been glimpsed shopping in Whole Foods, let alone drives a Prius"), "Killing Birds to Save the Planet" (wind energy conversion units as Cuisinart blenders), "Language" (a brief Orwellian guide to popular eco-verbiage), "Seuss, Dr." ("his doctorate may have been faked"), and so many more quotable gems.

Perhaps the best spot-on insight of all is under the heading "You Didn't Hunt for the Missing Children." Mr. Delingpole's explanation (borrowed from his friend Gary) of why, "[w]hen the current global warming non-crisis blows over," no credit "will be given to the minority of people who were right by the majority of people who were wrong," is alone worth the price of the book.

An Unforeseen Climate Crisis

A CLIMATE CRISIS of global proportions is unfolding right before our eyes and even the world's most powerful leaders can't do anything to prevent it. The coming years may end up being a period of declining global temperatures at precisely a time when carbon dioxide levels continue to rise—the opposite of what was predicted by the vaunted climate models.

Man the barricades! Something must be done immediately. The options are: 1) revise the selection and interpretation of data to vindicate the climate visionaries, 2) issue some scientific sounding mumbo jumbo to explain away the climate's failure to conform to the consensus view, or 3) simply ignore reality in the hope that few people will want to admit that they've been had. Or we could try something a bit more theatrical: announce that Mother Nature has decided to give us a reprieve so that we can get our collective global act together before she really lowers the boom. After all, warning that disaster lurks just around the corner has always worked well in the past.

Whatever strategy is selected, it won't include an apology to skeptics or even an acknowledgement that they were right. More likely, skeptics will be blamed for the confusion.

Still, you have to wonder why the purportedly smartest people have not learned the simple truth that a forecast is just a guess about the future based on information about the past and present. It's been nearly 200 years since Pierre-Simon Laplace boasted that if he knew the location and momentum of every atom in the universe, then he could predict the future. Subsequent discoveries show that Laplace's deterministic view of the universe was naively optimistic.[44] Yet today many climate scientists believe that given enough temperature data they can predict the future of the Earth's climate. Worse, they have discovered that the more confidence they express about their prognostications, the more people are likely to believe them. The whole thing takes on a life of its own as people gravitate toward those who claim to have the answers.

When dealing with something as complex as the Earth's climate, no one can know the future with a high degree of certainty, and to imagine otherwise is the height of arrogance. It's the old confidence game: first they get you to believe they know what they are doing, and then they get you to believe they have the answers. The public is easily impressed by sophisticated climate models, and enough people will step forward to keep the game going. But the only people who win a rigged game are the ones doing the rigging.

The average person knows that even short-term weather forecasts can be unreliable. AGW proponents counter that weather and climate are not the same thing. But they are related. However, let's concede the point and consider short-term *climate* forecasts. The official U.S. forecast for the 2006 hurricane season is a great example. Arguably the world's best hurricane forecasters tried to predict the total number of severe storm events just prior to the onset of the Atlantic hurricane season in May of 2006. The official forecast stated it would be another season of unusually numerous events (though not on a par with the record-breaking 31 events of 2005, which included 15 hurricanes). However, the forecast was a complete bust, because only 10 events (5 hurricanes and 5 tropical storms) were recorded. Keep in mind that in the average year there are about 15 events. The point is simply this: forecasting short-term climatic conditions is as difficult as, if not more difficult than, forecasting next weekend's weather.

If we can't accurately predict the number of severe storm events for a portion of the globe over the short-term, then what are we to make of the substantially more complicated and difficult task of long-range global climate forecasting? Like it or not, the Earth's climate doesn't play by the climate wizards' rules. Perhaps it's time to admit that we're not as smart as we think we are—that the Earth's climate is tremendously complex, and though we understand it better today than we did several decades ago, we are still a long way from being able to accurately forecast climate conditions decades into the future.

Moderating temperatures during the last decade have created a global crisis if only because some very powerful people met at a UN climate conference in Paris in 2015 to remedy predicted *increasing* temperatures. Luckily for them, recent months have been unusually warm. However, they've never let data that casts doubt on global warming get in the way. Perhaps they are less concerned about understanding the Earth's ever-changing climate and more concerned about redistributing the planet's wealth.

The Taking of Climate Science 1 2 3

REGARDLESS OF THE ultimate impact of Climategate, several crucial facts about the atmosphere surrounding the climate change debate have long been apparent. The e-mail messages and documents from the University of East Anglia's Climate Research Unit (CRU) exposed in the fall of 2009 and again in the fall of 2011 show that self-appointed gatekeepers have been limiting the information about climate change shared with the public.

Three important facts have been purposely obscured by the presumed experts:

It has never been established that there is a consensus among the relevant scientists that humans are causing long-term, global climate change. (However, see telling results from a recent survey conducted among members of at least one pertinent scientific society—the American Meteorological Society (AMS).[45])

Climate models have always been crude tools for estimating the long-term global consequences of various trends within the Earth's extremely complex climate system.

Water, not carbon dioxide (CO_2) or any other greenhouse gas, has always been the most significant climate regulator.

Let's take a closer look at the much ballyhooed consensus among scientists that human activity is causing dangerous global warming. Climate czar Carol Browner (she was Director of the White House Office of Energy and Climate Change Policy from 2009 to 2011) said "I'm sticking with the 2,500 scientists" who reviewed the U.N. IPCC's documents ascribing global climate change to carbon emissions produced by humans. However, not all of the 2,500 expert reviewers are climate scientists. Even according to the Union of Concerned Scientists, the IPCC reviewers are not all atmospheric scientists. Industry representatives, NGO (non-governmental organization) experts, and others participate in the process. And IPCC reports are only released after they are approved by government representatives.[46]

Are the IPCC's reports accurate, reliable, and fair? The IPCC and its supporters tell us that they are. For example, the Union of Concerned Scientists has this to say about the IPCC review process: "Many authors attest that this review process ranks among the most extensive for any scientific document." It's nice that "many" of the IPCC's authors attest to the quality of the IPCC's review process, but isn't that rather self-congratulatory? The IPCC and its supporters make many claims about "experts" and "detailed rules and procedures," but they don't see the need for a thoroughly independent assessment.[47]

Prior to the AMS survey identified above, little was done to find out what the thousands of scientists not involved with the IPCC but doing climate-related work think about the hypothesis that humans are substantially responsible for global climate change. Furthermore, the IPCC describes its role as "...to assess on a comprehensive, objective, *open and transparent basis* the scientific, technical and socio-economic information relevant to understanding the scientific basis of risk of *human-induced* climate change, its potential impacts and options for adaptation and mitigation..." [emphasis added]. The only evidence that the IPCC conducts its business in an "open and transparent" manner is the fact that it holds numerous meetings attended by like-minded people. More revealing is the IPCC's admission that its role is "understanding the scientific basis of risk of *human-induced* climate change," which clearly assumes that human-induced climate change is real. That doesn't leave much room for investigating alternative hypotheses. So much for claims that the IPCC is inclusive and fair.

Though climate models are useful for researching and understanding the dynamics of the atmosphere, they remain relatively primitive. And it's certainly not reasonable to rely on today's climate models to predict long-term global climate changes with any specificity. Again, clouds play a profound role in regulating the earth's surface temperature, but they are not adequately represented in computer simulations, and that affects the accuracy of long-term global climate forecasts.

Which brings us back to water's role as a key climate regulator. As mentioned earlier, the global atmospheric temperature is largely controlled by water in its various forms. Governments, however, would have trouble regulating water based on the claim that it's a dangerous pollutant. It's much easier to pitch the idea that CO_2 is bad for polar bears, because most people don't understand the vital role that CO_2 plays in plant and animal life.

Besides, the hypothesis that CO_2 causes global warming is not definitively supported. Deep ice core samples representing conditions going back hundreds of thousands of years show global temperature increases occurring about the same time or even *before* global CO_2 increases occurred. In other words, the evidence suggests that temperatures may go up first and then CO_2 levels rise.

There are clearly good scientific reasons to be skeptical about climate change. However, the Climategate scandal showed that some scientists were manipulating and suppressing evidence, and otherwise doing everything they could to ostracize global warming skeptics. The explanations offered in defense of the CRU researchers are that the messages were taken out of context and that they merely reflect the kind of honest exchange that happens all of time. Apparently, we are supposed to believe that manipulating evidence, deleting emails, fantasizing about doing violence to opponents, and excluding skeptics from the peer review process are activities that all scientists engage in.[48]

We need to face facts. The AGW train departed the station long ago and flagging it down won't be easy. Proponents are not interested in debating skeptics or considering reasonable alternative tracks. But thanks to the Climategate scandal, the public has gotten a glimpse of some very rude tactics—tactics that have no place in modern science.

SideBar18: Climate Change Conclusions: You Get What You Pay For

With the September 2013 release of the Summary for Policymakers of the first volume of the fifth Assessment Report (AR5) of scientific evidence behind climate change by the U.N.'s Intergovernmental Panel on Climate Change, it was time to recall the IPCC's original stated purpose. In their own words, the role of the IPCC is to assess the "risk of human-induced climate change, its potential impacts and options for adaptation and mitigation..." In other words, the organization's mission assumes from the get-go that anthropogenic global warming is a fact.

So, it is understandable that a climate researcher looking for financial support will craft their study proposal in such a way as to meet the needs of public and private entities seeking to endorse the IPCC position. This is not dishonest. The researcher is meeting a need and being paid to meet that need. Fair enough.

But, "you get what you pay for" and this is not how authentic scientific research is supposed to work. As pointed out by Al Gore himself in his movie An Inconvenient Truth, socialist Upton

Sinclair observed that "It is difficult to get a man to understand something when his salary depends upon him not understanding it."

Consider the alternative. Suppose an organization pays someone to research the possibility that human impact on long-term, global climate change is negligible. Is this considered improper with respect to authentic science while payment-to-endorse-human-induced-climate-change is not?

Furthermore, authentic scientific research requires that a hypothesis like "humans are responsible for long-term global climate change" be falsifiable; that is, able to be disproven. And, if it is disproven by, for instance, prognostications not matching reality, then the hypothesis should be discarded or reworked.

For more than 18 years now, as greenhouse gas emissions continue to rise sharply, global temperature increases have not materialized as confidently predicted. So, understandably, what we now have by those championing the science-for-hire model is a frantic rush to continue to defend the hypothesis that humans are largely responsible for long-term global climate change via "carbon pollution" emissions.

Frenetics are required since too much is at stake; too much money, effort, and prestige have been spent spinning-up a state of fear [h/t to the late Michael Crichton].

Yet, too much is truly at stake for scientific practice. The climate of contemporary scientific research must follow a more objective course, so that such research is in the service of humanity, rather than servant to the highest bidder.

The Copenhagen Cabal

THE CLIMATEGATE SCANDAL first broke in November of 2009. Predictably, the UN's climate change summit in Copenhagen in December 2009 was marked both by blustery weather and blustery rhetoric.

Unfortunately, the unseemly practices of the Climategate scientists did not foreclose business-as-usual at the summit. Too much money is riding on human-induced global warming to let a little deception and intimidation gum up the works.

You have to admire the artistry with which the Climategate e-mail authors and their many supporters in government, academia, and the media averted disaster. With their credibility seemingly in tatters, any other group of mere mortals would have been hobbled for a long time. In this case, however, despicable comments were simply recast as a simple misunderstanding about how elite scientists chat among themselves. The eavesdroppers were the real wrongdoers. And shame on the critics for misinterpreting their lofty conversations. Be thankful there are scientists who know how to add value to raw climate data.

So with that little misunderstanding cleared up, progressive scientists could get back to the urgent tasks of identifying and remedying the Earth's most pressing problems. The world's poor were particularly lucky to have such sapient and capable negotiators making deals on their behalf in Copenhagen amidst the opulence that is typical of such gatherings.

A reasonable and compassionate person could have easily been confused by the Copenhagen climate change circus. Aid was ostensibly being offered to the planet and its poorest inhabitants. But as frequently happens, the delegates of doom seemed to ignore the very terra firma and languid populace they claim to champion. They focused on cutting carbon emissions, which may accomplish little or nothing in the way of climate change. However, because slashing carbon emissions limits use of the most readily accessible and inexpensive energy sources, it hampers the ability of developing nations and the poor in general to climb out of poverty.

There are several ways for a compassionate person to make sense out of these summits. Perhaps some of the delegates sincerely believe the Earth is in deep trouble and our only hope is to cancel the Industrial Revolution. Other delegates may simply want to stop the spread of industrialization based on the happy thought that those who are not yet addicted to modern conveniences are less likely to miss them. The third explanation is more disturbing: Could it be that there is a tremendous amount of power and money being redistributed at these climate change confabs?

Even religious people at the Copenhagen summit seemed hell bent on advancing the vaunted wishes of their Mother rather than the humanity enriching and liberating desires of their reputed Father. They acted as if the Earth's fate was in their hands. They anointed themselves to nudge the data here and erase it there. They decided how to allocate the Earth's resources. And both secular and religious representatives arrogantly acted on the belief that they have both the power to destroy the Earth's climate and to save it from destruction.

A reasonable and compassionate person should ask a very different question: If representatives from all over the globe are brought together to decide how trillions of dollars' worth of resources will be developed and allocated, then why not focus on saving humanity? The list of needs is long: quality medical care, liberation from political oppression and brutal terrorism, sufficient drinking water, and clean air (removing genuine contaminants, not carbon dioxide)—to name a few. These are palpable and urgent needs, particularly among the world's poor. But we are told to put these issues aside and concentrate on events that may occur 50 or 100 years from now. That seems cruel given that without real help *now* some of world's poor won't even make it to next year.

The implied goal of Copenhagen and subsequent UN climate conferences is the establishment of a world government. In a blog post entitled "Effective World Government Will Be Needed to Stave Off Climate Catastrophe," [49] *Scientific American* editor Gary Stix amplified sentiments expressed in a policy article in *Science*.[50] Scientists, much like Hollywood celebrities, believe that success in their own field qualifies them to run the world.

However, it's easier to fall for utopian schemes than to undo the damage that they cause. A global government that transfers wealth from rich countries to poor countries must sound good to some people, but it's one of those schemes in which we all end up losers. Rather than making everyone comfortably equal, such redistributionist fantasies inevitably produce a new ruling class— the people who know best how to exploit hysteria, fanaticism, and naiveté. They make the middle class feel guilty and then ask them to cut back, do without, and share what little they have. The result is always the same: the ruling class gets rich, the middle class shrinks, and the poor get nothing.

Despite the sanctimonious words and deeds, the Copenhagen cabal is neither reasonable nor compassionate.

Checklist Climate Science

MANY OF LIFE'S mundane chores can be relegated to simple checklists: bread, milk, eggs... stop mail, leave lights on, lock doors... check, check, check. However, should simple checklists be used to address issues affecting the public—particularly when they involve complex scientific problems? Societal challenges often require carefully thought out solutions involving compromises that defy simple categorization. Therefore, when simple checklist items are offered as solutions to complex societal problems, it may signal the influence of a political ideology with its telltale preconceived notions. And wouldn't you know it: simple checklist solutions are often offered when the public is confronted with environmental and energy policy issues.

The observation and analysis of atmospheric variables such as temperature, precipitation, cloud cover, and wind data is a tedious process that should be conducted in a dispassionate manner. Unfortunately, the assessment of climate data is often tainted by the foregone conclusion that the Earth's atmosphere is changing due to man's reliance on fossil fuels. Belief in anthropogenic global warming has become a litmus test for "real science."

However, as physicist Richard Feynman said at the National Science Teachers Association's annual meeting in 1966, "Science is the belief in the ignorance of experts." [51]

Unfortunately, in the current political environment, the goal

123

of scientific observation and experimentation is often to confirm that industrial activity is distorting natural climate patterns. Specifically, research is frequently designed to affirm the following checklist items:

Fossil fuels are harmful—Coal, oil, and natural-gas combustion produce emissions—primarily carbon dioxide—that directly or indirectly disrupt the Earth's climate. Natural forces are easily overwhelmed by the combustion of fossil fuels.

Humans are endangering the planet—Though natural forces and corrective human action may temporarily alleviate the trouble, the overall effect of human activity is pernicious.

Action is required immediately—The world community has very little time to act. The longer we wait, and the longer we tolerate harmful human activity, the longer it will take to properly address the problem and avert catastrophe.

Check, check, check.

It's a brief but all too familiar list. After all, human-induced climate change has been blamed for everything from A to Z: from the increased risk of an Asteroid strike to Zebra mussel infestations.[52] Supposedly, the risk of an asteroid strike is increased because global warming causes the Earth's atmosphere to expand, creating a larger target.[53] And zebra mussels are attracted to warmer waters.[54]

Name a real or imagined environmental problem, and human-induced climate change is likely to be cited as a cause—if not *the* cause.

That's not surprising. If the IPCC's mission is, "understanding the scientific basis of risk of human-induced climate change, its potential impacts and options for adaptation and mitigation," then there is money to be made doing just that. Research grant seekers merely need to cite the importance that a major UN agency attaches to such matters to boost their chances of obtaining funding.

Fame, fortune, job security… Check, check, check.

Science for hire often produces dubious results. Researchers pick topics that are most likely to attract government grants. Studies are commissioned to support pet theories. Unless there is a determined effort to let the research chips fall where they may, the truth may end up being ignored or even suppressed.

And when the accuracy and reliability of politicized research is questioned, grant recipients and their supporters circle the wagons. For example, when the Heartland Institute held its 2011 Conference on Climate Change featuring skeptical atmospheric scientists,[55] the Center For American Progress responded with a telephone conference for the press to discuss "the recent extreme weather, the science behind climate change, and the flawed arguments of those refuting its manmade nature."[56] However, the announcement about the telephone conference seemed less concerned about science and more concerned that the Heartland Institute's "sponsors in years past have included groups backed by Exxon Mobil and the Koch brothers." Mind you, the sponsors weren't Exxon Mobile and the Koch brothers, but groups *backed* by them.[57]

Though checklists are terrific for remembering grocery items and preparing for a vacation, they're terrible for resolving scientific controversies and planning national energy policy.

Intellectual bullies love checklists, because checklists help them frame the issues the way they want them framed. They prey on people's fears (that future generations will suffer), sympathy for the little guy (against Big Oil), and insecurity (vis-à-vis proclaimed experts). But it isn't helping global warming activists. Gallup's Environment poll[58] conducted in early March 2011 revealed that "Americans continue to express less concern about global warming than they have in the past..." even as their "self-professed understanding of global warming has increased over time—from 69%...in 2001, to 74% in 2006 and 80% in the current poll." However, with financial support from Big

Government and additional support from Big Media and Big Education, the purveyors of manmade climate catastrophe don't need the public's support. They just need the public's acquiescence.

Convincing the American public that their use of fossil fuels is pushing global temperatures dangerously high is not going to get any easier. More and more contrarian views and exposés are being published, such as the book by University of Houston professor Larry Bell, *Climate of Corruption: Politics and Power Behind the Global Warming Hoax*[59].

Bell's work, which serves as a sort of alternative checklist, is a good overview of government-sponsored climate research and findings for beginners. It is also a good synopsis for atmospheric science veterans of the latest fads and fallacies emanating from the global warming industry.

Bell guides readers through the history, hysteria, politics and power behind what he characterizes as a hoax. In the first section, "Setting the Records Straight," Bell looks at past and recent climate history, noting that a "basic tactic used by calculating 'hysteria hypesters' is to treat propaganda as obvious fact." In section two, "Political Hijackers of Science," Bell continues the theme that "[i]t is difficult to imagine a time in recent history when so much political hype has swirled around so little substance." He goes on to ask a question many others have asked, "Is it logical to wager trillions of dollars based upon flawed science practices and suspect agendas?" Section three, "Carbon Demonization Scams," looks at cap-and-trade, climate science as religion, and "green energy."*Climate of Corruption* succinctly summarizes the scientific and political problems that are created when science is politicized—particularly how doing so jeopardizes national and global energy supplies.

Practical solutions are clearly spelled out in section four, "Retaking America's Future." Bell proposes renewing free enterprise, demanding truth and accountability in scientific practice, and acknowledging and practicing American exceptionalism—relying on the ingenuity of the American people to solve problems.

Throughout *Climate of Corruption*, Bell demonstrates that there are "big differences between environmental stewardship ideals, which most of us subscribe to, and the ideologically moralistic, antidevelopment, obstructionist activism that exemplifies much of today's environmental zealotry." In other words, it's progressive ideology that misguides efforts to identify and address real threats to the environment, energy supply, and the common good for all of humanity.

The problem isn't that checklists are inherently bad. For example, imagine checklist items such as "implement sensible pollution safeguards" and "design solutions that balance the interests of industry and the public." The problem occurs when activists try to impose simple, one-size-fits-all solutions where compromise is required. It's good to see that despite checklist climate science, relentless global warming propaganda, and massive natural disasters that the American public is maintaining a balanced perspective. In part, that's because books such as *Climate of Corruption* are making the case that redistributing wealth is neither morally right nor effective. If we want to prevent or at least alleviate environmental disasters, then we need to leave space for American ingenuity, volunteerism, and local solutions.

SideBar20: 10 Reasons Climate-Change Hysterics Continue

With the global temperature trend still rather flat after 18 years and perhaps on the decline, why does climate-change hysteria still have a pulse? That's easy to answer: Everybody wins!

Here are 10 winning reasons for continued climate-change hysterics:

1. Indoctrination from grade school through graduate school has inculcated the "incontrovertible conclusion" that people are destroying the planet. By acting to save the earth, precious self-esteem is elevated, while guilt is assuaged.

2. Lack of depth of understanding about science and scientific practice, not only because of being uninitiated, but partially because inadequate science education has left the public either clueless about, intimidated by, or apathetic to science in general and climate change in particular.

3. Man-made climate-change hype acculturation has infused acceptance of human culpability into the psyche of everyone, from industrialists and businesspersons to the "man on the street."

4. Literally billions of dollars are up for grabs with consultants making beaucoup bucks advising on carbon credits, technocrats raking in the cash with carbon dioxide control and sequestration contraptions, and researchers securing grant money to tie every wind of change to human excesses.

5. Those who sincerely believe they know the long-term future of the global climate are committed to the cause. Commitment can be admirable, but nobody, no matter how smart, can predict the future climate decades ahead with any serious degree of accuracy. That has already been demonstrated with the leveled temperature trend that belies predictions.

6. Politicians and bureaucrats can increase their power over people. Control over energy is near ultimate control.

7. Journalists and bloggers have found a juicy, fruitful topic to squeeze.

8. Environmental and social activists have discovered a new "higher-calling" cause to champion and cash in on.

9. Sales of T-shirts and bumper-stickers advertising imminent world environmental cataclysm and its simple solutions—"Go Green," "Hug a Tree," "Love your Mother (Earth)," "Death to Deniers" (I just made up that last one, I hope!)—would dry up

like the Aral Sea. Without such capitalistic merchandizing where would socialism be?

And last, but certainly not least:

10. People get to defend their deeply held religious beliefs and can feel they're doing something good for Jesus, God, the Buddha, Vishnu, Gaia, the Universe, children or grandchildren, pets, polar bears, plankton.

So, everybody wins... everybody that matters, that is, but not the middle-class who ultimately end up footing the bill, and definitely not the poor who are simply used as a sanctimonious diversion, yet end up as impoverished as ever.

The Earth is Most Definitely Doomed—Maybe

IN THE PREFACE to his latest book, *A Universe from Nothing: Why There is Something Rather Than Nothing*,[60] theoretical physicist Lawrence M. Krauss states:

> Science has been effective at furthering our understanding of nature because the scientific ethos is based on three key principles: (1) follow the evidence wherever it leads; (2) if one has a theory, one needs to be willing to try to prove it wrong as much as one tries to prove that it is right; (3) the ultimate arbiter of truth is experiment, not the comfort one derives from one's a priori beliefs, nor the beauty or elegance one ascribes to one's theoretical models. (p. xvi.)

The climate scientists who insist that human activity is causing dangerous global warming violate all three principles. They don't follow the evidence, they lead it—which is easy to do when your evidence consists primarily of proxy data input and computer model output. Rather than try to prove their own theory wrong, they discredit and silence anyone who *does* try to prove their theory wrong. Perhaps most revealing, they clearly trust their a priori beliefs, because there cannot be verifiable and repeatable experiments to test predictions about the distant future.

Many prominent climate scientists say that if humans continue to burn fossil fuels then the Earth is definitely headed for a calamity. Anyone who disagrees with this prognosis, whether or not they are climate scientists, and no matter how knowledgeable, skilled and experienced they are, is labeled a "climate change denier" and lumped together with every sort of crank.

This name calling is unscientific, unfair, and unprofessional. Science is supposed to be based on facts and logic. But ad hominem arguments are based on a logical fallacy—that ideas can be discredited by attacking the person advocating those ideas. People who resort to name calling are no more scientific than people who are guided by superstition.

Name calling is unfair, because it distracts people from considering the merit of what opponents are saying. Calling AGW skeptics "climate change deniers" is particularly disingenuous; none of the atmospheric, earth, and environmental scientists that I know denies that climate changes. Rather, they are unconvinced[61] that human activity is causing harmful long-term, global climate change.

Name calling is also a desperate and unprofessional tactic. Even the best scientists cannot know with certainty what is going to happen in the distant future. Yet many climate scientists claim that they do know. It's a big problem for them when critics bring up the knotty and obstinate laws of physics, chemistry, biology, and statistics. That's quite a bit to overcome. It saves a great deal of time and effort to simply dismiss such critics by smearing them.

The flipside of smearing opponents is snowing the public. Sophisticated, computer-based climate models make climate fortune-telling appear exacting, rigorous, and highly scientific. However, models of complex processes are merely tentative representations based on someone's interpretation of the available data. There are good reasons for using such models, but there are also good reasons to treat the results with caution.

First, models are merely representations—and tentative ones

at that. Scientists don't typically view a model as a miniature version of the real thing. Rather, models represent reality via mathematical equations that simulate what scientists have observed. However, models are not truth generators. In fact, according to ASTM International (formerly known as the American Society for Testing and Materials) "verification of the truth of any model is an impossible task." [62]

Next, keep in mind that our observations are limited by the data that is available, the data that we choose to gather, and the accuracy of our measurements. If the data is scarce, incomplete, or inaccurate, then that will affect the reliability of the mathematical equations used in our computer models. Or as computer technicians used to say during the batch processing days, "Garbage in, garbage out."

Finally, even if our information is reasonably complete, it still has to be interpreted. This is where bias can rear its ugly head. Preconceptions and desired outcomes are a constant threat to objectivity. The truth suffers when modelers, whether consciously or more often unconsciously, select inputs to their models that produce desired results. Some of the dubious practices that can distort computer models include the selective use of "random samples" of data and the tweaking of algorithms to produce less ambivalent results.

Fortunately, computer models can be tested by seeing how well they predict past events. Predictive ability establishes the model's usefulness. If a model is a reliable and accurate predictor of past events, then there is reason to believe it has some ability to predict the future. However, to get a model to perform well—to make it produce accurate predictions—it's usually necessary to fine-tune the model. While fine-tuning may produce a better match to the past and the near future, the same cannot be said when peering farther into the future. The predictive powers of climate models decrease inversely with time: the farther into the future you try to predict, the less reliable the prediction.

It stands to reason that by collecting more and better atmospheric data, and by properly interpreting that data, better atmospheric models can be created. The lack of sufficient, good quality input data is a serious problem, however, and it's currently being addressed by the scientific community. Given that our atmosphere is like a vast and complex ocean, it's understandable that there is still a paucity of observed data for crucial meteorological variables. But even if the quantity and quality of the data is improved, faulty interpretation of the data is no small mistake—particularly when interpretation is influenced by overconfidence in planetary projections by self-proclaimed climate gurus and crisis-mongering politicians.

Scientists are only human. They tend to believe what they want to believe. They seek validation of their beliefs through the approval of other people. And they are tempted to seek validation of their beliefs in the data. That's why science demands that investigators try as hard to falsify their propositions as to confirm them (per Lawrence Krauss' second principle, above). The best way for scientists to discover the truth about nature is to submit their theories to the harshest testing—to question their own assumptions and invite doubters to tear apart their conclusions.

Manmade global warming activists say that doing nothing is too risky. But spending a trillion dollars or more to "solve problems that don't exist with solutions that don't work"[63] is also risky. If we act now and discover later that the long range predictions of imperfect computer models were wrong, it won't be just the science establishment and several oversized egos that take a beating. There will be many losers and they will include the industries that were punished for using efficient technologies, the people who were rewarded for using inefficient and unreliable technologies, and the people who could have been lifted out of poverty had the money been better spent.

However, science may take the biggest hit for allowing itself to be politicized. Because when science deals in compromised evidence, settles for consensus opinions rather than experimental proof, and discourages criticism of popular theories, it ceases to be science.

SideBar21: Everybody Knows That about Climate Change

With a (very small) hat tip to that cute little GEICO gecko, here's a possible real-life conversation...

The Scene: Curious reader pondering headline in local paper.

"Huh...97% of Homeopathic doctors believe homeopathic medicine works?"

"Yeah, everybody knows that."

"Well, did you know that left-wing ideologues make lousy climate forecasters?"

"What?"

"Sure, scientists saturated with a leftist education from grade school through graduate school have predicted for years that the global climate should now be a few fractions of a degree Celsius warmer. Yet, the globe has not shown any statistically significant temperature rise for at least 15 years."

Cue the Announcer: "A switch to real-world science, rather than that ideologically-motivated stuff...could have saved you 15 years or more on hysterical global-warming assurance."

Fixing What Ain't Broke

AS THE SAYING goes: "If it ain't broke, don't fix it."

Lately there have been many proposals[64] for fixing our supposedly broken global climate. Luckily, though one group of humans broke the global climate, there is another group of humans who can fix it.

One of the most popular among a crateful of recommended fixes calls for injecting sulfates into the lower stratosphere to simulate a volcanic dust veil—much like the one produced by the eruption of Mt. Pinatubo in the Summer of 1991. The dust particles ejected into the atmosphere by Mt. Pinatubo caused global temperatures to drop about one degree Fahrenheit for more than one year. If Mother Nature can cool the entire planet just by spewing dust into the atmosphere, the thinking goes, then surely humanity can do the same.

Another proposed fix employs the time-tested art of cloud seeding. Seeding can be used to increase cloud cover or brighten clouds—either of which will increase the amount of sunlight reflected back into space, cooling the planet in the process.

Other solutions focus on reducing carbon dioxide levels. For example, one idea is to fertilize the oceans with iron, coaxing plankton to gobble up more CO_2.

Another way to reduce carbon dioxide levels is to deploy artificial trees that extract CO_2 from wind. The CO_2 would be routed underground where it would be stored as a solid or it could be injected deep into the nearest ocean. Tens of thousands of these devices would be needed to absorb all the CO_2 spewed into the atmosphere by the US alone. The estimated cost of this environmental engineering scheme would be nearly $600 billion per year.

One thing that's certain: many scientists and engineers would benefit financially from technical solutions to anthropogenic climate change. And they wouldn't be alone. There would also be opportunities for software engineers (to develop programs for tracking and reporting carbon emissions), economists, cap-and-trade brokers, entrepreneurs, and investors. But the biggest beneficiaries would probably be politicians and bureaucrats (who would manage the whole thing) and attorneys (who would sort out disputes).

All of this assumes that humans are causing large-scale climate change and, therefore, that humans are ethically responsible for fixing it. The same human ingenuity that accidentally heated the planet could be harnessed to cool the planet.

However, what if there is long-term global climate change but it is not primarily due to human activity? Would it still be wise to intervene in the Earth's climate affairs? Or would the best strategy be to simply accept the Earth's natural variability and adapt to it as best we can? After all, if climate change is a natural process and we disturb it, isn't there a possibility we will unintentionally make things worse?

Once we embark on fixing the climate will we know when to stop? And will we be able to stop? Given all of the money flowing from individual and corporate taxpayers into government agencies, and all of the money flowing out to researchers and contractors, won't there be incredible pressure to perpetuate the climate change crisis?

What if the whole thing turned out to be unnecessary? What if the climate gurus were forced to give the data a second look and realized that they are not as prescient as they once imagined?

After all, we've been down this road before. In the 1970s it was widely believed that plunging global temperatures were causing the glaciers to advance, and proposals were submitted to reverse the process. If the soot solution described in Chapter 2 had actually been implemented, then where would we be today? Perhaps we would be a little warmer. But mainly we would be trying to figure what to do with truckloads of dirty snow.

Whether humans are culpable for climate change or not, many people are staking out claims for the anticipated gaseous gold rush. That reminds me of one of my favorite proposals for countering global warming: turn people into mobile carbon sequestration units.[65] And make no mistake about it some climate-fixers are poised to reap billions of dollars from the imposition of cap-and-trade regulations. While a relatively small and elite group will live in uninterrupted opulence, I fear a much larger number of people will be left to suffer at the bottom of the heap. And it won't make much difference to them whether they sweat or shiver.

SideBar22: Norman Borlaug and Human Ingenuity

Around the turn of the nineteenth century, a well-respected scholar and political economist proposed that, in nature, as populations increase in size geometrically, the populations eventually outstrip their food supplies, which only increase arithmetically. The concept was and still is employed to argue "limits to growth," especially in human populations. No wonder environmentalists believe the number one environmental hazard is a burgeoning global population!

The problem is that the population-to-food conundrum—proposed in 1799 by Thomas Robert Malthus—was soon roundly routed by technology, a technology used to great advantage many times since the early 1800s.

One person especially astute at using technology—in particular, genetic engineering—to feed many souls was Norman Borlaug. Mr. Borlaug was a devout Christian whose desire to help the needy was put into practice by starting a true "Green Revolution" in agriculture. His impressive accomplishments included developing genetically unique strains of wheat and rice to raise crop yields by as much as six-fold.

A few days after Mr. Borlaug died at the age of 95, the *Wall Street Journal* editors commented on September 13, 2009, that the 1970 Nobel Peace Prize winner "showed that nature is no match for human ingenuity in setting the real limits to growth. ... Borlaug showed that a genuine green movement doesn't pit man against the Earth, but rather applies human intelligence to exploit the Earth's resources to improve life for everyone."

Today, more than two hundred years since Reverend Malthus staked his mistaken claim and mere decades since Mr. Borlaug dramatically repudiated Malthus, people everywhere are in *potentially* better standing than ever to take care of their neighbors around the world. "Potentially" is the key word, because inept and oppressive governments and misguided policies are on the list of reasons why, even with food supplies worldwide sufficient to feed the global population for well over the past century, so many are still severely malnourished and even starving to death.

Climate Change Weather

WHEN THE LONG-AWAITED spring thaw finally came to the northeastern states in 2010, it was time to reflect on the recent weather. Not just the weather events, but their relationship to assumed human-induced climate change and the expensive remedies prescribed to rescue the planet from meteorological mayhem.

Weather is a pawn in the climate change game. When temperatures are unusually high for several days in the Northeast, activists chalk it up to global warming. When snow falls for days in the mid-Atlantic region, skeptics cry "Foul!" What does local weather have to do with global climate change? Quite a bit. It's the day-to-day manifestations of atmospheric dynamics in locations all over the planet that define the global climate. And climate conditions are definitely changing—locally, regionally, nationally, and internationally. So what else is new?

To understand climate change, we need to look beyond weather events.

When temperatures rise, the atmosphere is able to retain more water vapor. Some people use this simple fact to conclude that heavy snowfalls are just further proof of global warming. However, it takes more than water vapor retention to produce snowfalls measured with yardsticks, bursting thermometers, and hurricanes named Hannibal.

Think of the atmosphere as an ocean. Thermal differences,

circulatory patterns, chemical additions and subtractions, and other factors conspire to make the airy seas stir, swirl, and swell. And like an ocean, the atmosphere has the ability to mitigate large-scale disturbances and maintain its overall character. In fact, you could even say that the atmosphere has a system of checks and balances that works to maintain the status quo. Yes, there may be dramatic twists and turns along the way, but these typically arise from isolated events such as (in the ocean's case) tsunami-producing underwater earthquakes or massive ice sheets that slide into oceans, raising sea levels. However, there have been tidal waves and changes in sea level throughout history and no one argues that they were caused by humans.

Water vapor, as well as ocean water and ice, are the key components of the atmospheric system that keep global average temperatures bobbing around 59 degrees Fahrenheit. An increase in water vapor can slow and even reverse rising temperatures by producing more clouds and occasional intense storms. In other words, much of what we observe in terms of weather events is just nature's way of making adjustments to stabilize thermometers—if not to calm people's fears.

General air circulation patterns are another major contributor to diurnal weather conditions. During the February 2010 "snowmaggedon" on the east coast, the southerly component of the storm track drew in huge quantities of Gulf and Atlantic sea moisture, enhancing the storm's snow producing capacity. And in more recent months, deep shifts in the mid-latitude jet stream (rivers of fast flowing air about 30,000 feet overhead) sent temperatures alternately soaring and sinking across portions of the US.

Alterations to air circulation patterns can sometimes be attributed to chemical changes in the atmosphere such as increased carbon dioxide levels, but they are more likely due to natural variations in parameters such as water vapor content, sea surface temperatures and circulation patterns, ground cover, and solar irradiation.

It's comforting to know that this natural system maintains tolerable living conditions, perhaps not everywhere and at all times, but on balance in enough places and times to keep most people satisfied. It's also nice to know that we have the resources and the will to mitigate the damage caused by natural disasters. However, it's quite doubtful that humans have exceeded the combined power of natural forces, and even more doubtful that humans are overpowering natural forces as an unintended consequence of industrial activity. Rather than trying to "correct" the Earth's climate based on proxy data and untested computer models, we should continue to study the Earth's climate and improve our ability to respond effectively to natural disasters.

To maximize our effectiveness, we need to have the correct priorities. That means putting our information into order: what we know, what we think we know, and what are just guesses. It also requires putting our choices into order: what we can do, what we think we can do, and what we might be able to do. If we want to do something about severe and changing weather, we can start by helping the victims rather than using them—like the weather is used—as pawns in the climate change game.[66]

SideBar23: Keeping Cool on Climate Change

It's been said that Mark Twain (or was it Samuel Clemens) came up with what is probably the best succinct distinction between climate and weather: "Climate is what you expect, weather is what you get."

But, with the early 2014 and 2015 frigid temperatures in the U.S. Midwest and East and icy conditions for much of the nation, it was important to keep a cool head and make the right division between climate and weather. Perspective and precision are essential to science, especially contemporary climate science. So, before anyone goes off making ridiculous unfounded claims that these conditions somehow had disproved the hard fact of man-made global warming, let me remind everyone of the contemporary difference between "weather" and "climate" (the Mr. Twain definitions not withstanding). James Delingpole of the *Daily Telegraph*, in his *Little Green Book of Eco-Fascism*, defined the difference best with:

> *If it's cold when the greenies say it should be warming, it's weather. If, on the other hand, it's doing anything that bears any correlation with their computer models' predictions of environmental doom, then it definitely counts as climate.*

Earth Day: Power to the People

THE 40TH ANNIVERSARY of Earth Day was celebrated on April 22, 2010. A product of the 1960s counterculture, Earth Day was established to hold teach-ins about the environment and to promote back-to-nature lifestyles. Participants, who include celebrities and politicians, have identified people as the planet's Public Enemy #1.

Prepare for even more hyperventilating because in 2011 Planet Earth became home to 7 billion human inhabitants. Anguished environmentalists, who put the planet before people, will continue their struggle to save the world by reducing excess human births. After all, people have been told for decades that overpopulation threatens our ability to feed everyone and maintain a healthy environment.

Is the planet really in trouble because of the number of human inhabitants, or has blaming people simply become a mantra of the environmental movement? What if we took a radically different approach to the supposedly adversarial relationship between people and the environment? First, however, we need to put the number of the planet's human inhabitants into perspective.

Imagine that we schedule a global population conference to which we invite every man, woman, and child—that's seven billion people—to one location. Let's allocate an 18 foot by 18 foot area—a nice size room—for each person. How much total space would we need to accommodate the world's entire population?

Believe it or not, we would only need an area the size of the state of Utah to give every person on Earth their own space. (See sidebar for the confirming calculation.)

There is room for many more people—particularly since people tend to congregate in cities, towns, and villages. And in many developed countries, population growth has come to a standstill.

The critical question becomes: If the sheer number of people on the planet does not threaten mass starvation and environmental destruction, then what does? The answer: a group interrelated factors that include *population density*, *politics*, and *individual choice*.

For example, the population density of Bangladesh is comparable to the population density of specific locations in the US such as Freemont, California. Yet living standards for the two groups are about as different as their geographical locations are far apart. What is the cause? It's *not* that there are too many people in Bangladesh. Rather, it's differences in natural and political climates. Bangladesh's natural resources and weather conditions can't sustain the number of people living there. And even if the harsh weather conditions could be ameliorated (as they have been in places such as Phoenix, Arizona and Las Vegas, Nevada), political mismanagement has kept Bangladesh's economy relatively undeveloped.

Individual choice may have the most substantial impact. Consider that one careless or malicious individual can cause tremendous ecological devastation (for example, a forest fire can be lit with a single match). Consider also the mayhem that can be caused by a group of people—such as anarchists or terrorists—who are hell-bent on destroying civil society.

Some argue that though there is enough space for the earth's 7 billion people, there isn't enough food to feed everyone. This is not a new argument. Around 1800, economist Thomas Malthus predicted that because food supplies tend to increase arithmetically over time while populations increase geometrically, the world would eventually experience mass starvation.

Though starvation on the scale predicted by Malthus has never occurred, his theory refuses to die. As recently as the 1960s and early 1970s, there was a spate of predictions of impending worldwide famine due to overpopulation causing tremendous civil unrest. Amazingly, the prophets of doom in those days included President Obama's current science and technology advisor, John Holdren. However, there has been enough food to feed the world's entire population for at least the past century. (While there have been local food shortages caused by wars and droughts, there have been recurring food surpluses elsewhere.) New technologies, unforeseen by Malthus and minified by Holdren and his friends, have helped to keep the world's food pantries well-stocked.

And despite the related mania over resource depletion, there are enough raw materials, including fossil fuel supplies, to serve the needs of everyone—presuming everyone is permitted access. Historically, shortages have been overcome through the development of new technologies and innovative products. For example, plastics serve as substitutes for many raw materials. New techniques and equipment make it economical to extract fossils fuels (for example, oil from bituminous sands) and minerals where previously it was not economical. And nuclear fuel can be used in place of fossil fuels.

The truth is that there are enough food and other resources to go around. Where there are shortages and widespread environmental damage, there are likely to be other contributing factors such as local weather, population density, politics, and personal choice.

All of this suggests that we should take a fundamentally different approach to problems such as starvation and damage to the environment. Namely, we should put the needs of people first.[67] Unfortunately, many environmentalists put the Earth first.

They want people to serve the Earth, rather than the Earth serving people. If we put people first and diligently use the resources that are abundantly available to sustain everyone, we will find that caring for the world's people naturally includes caring for the global environment. It's not necessary to punish people in order to save the Earth.

In other words, taking care of the world's people requires good stewardship of the Earth's bounty. Now there's an environmental approach worthy of the next Earth Day and all future Earth Days.

Good Friday and Earth Day: Freedom Versus Slavery on April 22, 2011

Two faiths clashed when Good Friday and Earth Day fell on the same date in 2011.[68] Friday, April 22, 2011 marked the solemn Christian holiday memorializing the crucifixion of Jesus, but this particular Friday also featured the 41st Earth Day. The values promoted by the two holidays could not be more dissimilar.

Christianity is based on the message and acts of Jesus, which focused on people. In addition to love and grace, freedom was one of his main evangels. This freedom was from all unjustified forms of servitude, which surely include the sacrosanct norms and standards of today's progressive environmentalism.

While many people are familiar with the structure and strictures of world religions, familiarity with environmentalism as a religion is not as common. Progressive environmentalism from its modern incarnation in 1962 with Rachel Carson's *Silent Spring* quickly emerged as a faith rivaling the world's major religions. It has all the trappings of a religion including a god (Mother Earth or Gaia); holy writs (*Silent Spring*, *The Population Bomb*); mantras ("Love your Mother," "Save the Planet"); doctrines (capitalism and industrialization are evil; socialism and eating low on the food chain are righteous); dead saints (John Muir, Rachel Carson); and, of course, holidays. In addition to Earth Day, there are the UN-sponsored World Environment Day on June 5 and World Ozone Day on September 5.

In practice, progressive environmentalism has by no means been a compassionate religion. The number of casualties and calamities resulting from its crusades has been almost incalculable. Elimination of the use of DDT to fight malaria-carrying mosquitoes alone has contributed to the deaths of millions of human beings in third world countries over recent decades. The diversion of US corn crops for the production of ethanol raised world corn prices and lead to riots across the globe in the spring of 2008.[69] In the US, restrictions on drilling ample reserves of oil and natural gas, mining abundant coal, and constructing nuclear power plants have resulted in national insecurity, worldwide instability, and even (as some assert) "war for oil." Environmentalism's culpability for worldwide misery ranks right up there with the most ruinous ideologies.

In fact, it is more accurate to compare environmentalism with medieval religions. In much of the West, environmentalism has become a state religion. It restricts people's daily activities. Environmentalism tolerates little or no dissent. And it is quick to brand any who dare to oppose it as "deniers," the new word for heretic.

Judeo-Christian civilization, meanwhile, evolved beyond medieval thinking to champion freedom of conscience, freedom of speech, and tolerance. It also gave birth to ideas including universal education, universal health care, and even conservation of our natural environment. Jewish and Christian philosophers led the way by demonstrating respect for facts[70] and logical discourse.[71]

No sensible, caring person denies that good stewardship is required for the use of natural resources and arable lands. However, progressive environmentalism has given rise to a careless denial of the sensible.

Despite the hysteria over resource depletion, and treating people as servants to the Earth as the progressive environmentalists would have us do, why not enlist the Earth in humane service to people?[72] The focus, as in Christianity, should be on people first, and their freedom to diligently use the resources that are abundantly available to sustain them. After all, people have the potential to do good and even great things. The Earth is not alive in the sense that plants, animals, and humans are. The Earth does not possess intelligence, free will, or a soul.

When the apostle Paul told followers, "It is for freedom that Christ liberated us, therefore stand firm and do not be burdened again to a yoke of slavery," he was referring to all forms of slavery. Such Good Friday advice should be embraced to break the bondage to beliefs others try to impose on us, especially many of the beliefs celebrated on Earth Day.

SideBar24: Putting People First on Earth Day

Radical environmentalists deem people as the top environmental problem. There are just too many people for our natural world to endure. We must drastically reduce the world's population at all costs. When the number of carbon polluting life-forms is brought back to a more sustainable level, then the planet will once again become a garden of delights in which all flora and fauna may frolic.

Utter nonsense.

On the 44th Earth Day (April 22, 2014), as today, there were about 7 billion people on planet Earth. What if we were to invite all these folks to one location on the globe, giving each individual their own 18 foot by 18 foot area (a nice-sized room) in which to celebrate Earth Day? How much total area would be needed to comfortably accommodate the world's population? A bit of simple math demonstrates that an area only the size of Utah would be quite sufficient![vii]

And don't worry about feeding all these people. There has been enough food to satisfy the entire population for at least the past 100 years. But, what about raw material and energy resources? Plenty of those too.

So, why the angst over too many people? The sheer number of global inhabitants is not the issue. More likely, environmental challenges stem from interrelated factors such as population density in topographically and climatologically unfavorable areas, ruling-class politics at odds with the needs of the populace it supposedly serves, and personal decisions to be wasteful and harmful to the planet and fellow inhabitants.

Therefore, what if we took a compassionately different approach to the purportedly adversarial relationship between people and the planet? What if we put the needs of people first, rather than the needs of the environment? If we put people first and diligently use the resources that are abundantly available to sustain every individual, we just may find that caring for the world's precious people will naturally necessitate caring for the global environment. In other words, responsibly caring for the world's people will require good stewardship of the Earth's bounty.

Now this is an environmental approach worthy of celebration, and not just on Earth Day.

[vii] Here's the calculation:

(7.16 x 10^9 people on Earth) (18 ft x 18ft space per person) = 2.32 x 10^{12} ft^2 required.

Area of Utah = 84,900 square miles = 2.37 x 10^{12} ft^2 available.

Help GAIA Save the Planet!

EARTH DAY, THE annual pagan holiday, has been celebrated with reverence and fervency for more than 40 years. But as the saying goes, "We set out to do good and ended up doing well." The Sierra Club, Environmental Defense Fund, Greenpeace, Natural Resources Defense Council, Friends of the Earth, and other non-profit organizations began small and are now multimillion dollar businesses.[73] During the early years, even with little financial support, these organizations wielded considerable influence and helped conservationists achieve great strides.

However, the environmentalists' organizations have changed—much like a business started in a garage that becomes a multi-national corporation. Today, many of these groups have enormous annual budgets. They have a significant presence on Capitol Hill and the political connections and lobbying experience to match. Environmentalist lobbying groups are distinguished from garden variety influence peddlers in that while most lobbyists represent clearly defined interest groups such as businesses and professions, environmental lobbyists claim to defend our habitat. The two sides are often depicted as pitted against each other: businesses lust for friendly legislation and huge contracts, while wholesome environmentalists merely long to protect and comfort Mother Earth.

Shouldn't environmentalists be as upfront about who they are against as they are clear about what they are for?

Oh well, if you can't beat 'em, join 'em. Inspired by their unassailable motives, and awestruck by Herculean efforts to save the planet flowing from the Durban, Copenhagen, Cancun, *ad infinitum* climate summits, we propose a new organization exclusively dedicated to combating the forces arrayed against Mother Earth. This new order, named after the Greek earth goddess Gaia, is the Group Against Industrial Activity (GAIA).

Here are some excerpts from our proposed brochure introducing GAIA to the world community: [viii]

An Urgent Need: Our planet has been ravaged for over a century by industry and (as of today) seven billion waste producers. We need cash, and we need it fast. We are serious and sincere, so you can rest assured your money will be spent wisely. Like all good not-for-profit organizations, we believe in sacrifice—yours. Your generous gifts will be used to fight mechanized agriculture, modern transportation, power plants, and pharmaceuticals—to name just a few.

Telling the Truth about Climate Change: Believe it or not, there are still a few individuals who deny that industry is poisoning and contaminating our world. Though few in number, their voices are amplified by talk radio and Fox News Channel. Fortunately, the consensus among good scientists and progressive politicians is clear: Those who doubt manmade global warming, whether they intend it or not, serve no better purpose than the few notorious individuals who deny The Holocaust. Help us shut down this modern day Flat Earth Society.

Climate Crisis Awareness: GAIA understands that it's not enough to be right; you have to win. In addition to helping our friends get elected to Congress and working with the United Nations to establish a world government, GAIA is committed to teaching our children how to tell real science from pseudoscience

---

[viii] Note to fanatical environmentalists: This chapter is satirical. Please do not send money to me or this fictitious organization.

and, with the help of the media, exposing the industrial menace. Working with the federal government and the UN, we plan to secure funding to have all delivery rooms, birthing suites, and freelance midwives supplied with DVD players cued up with An Inconvenient Truth ready to roll upon crowning.

Cap-and-Trade: GAIA uses proven scientific techniques to save the planet. For example, our models show that "cap and trade" rules will limit carbon emissions while creating a new source of revenues for environmental activists and allies. Using proxy data to set the caps, government revenue will be maximized, and government programs and agencies expanded. Help us make "green" history in more ways than one.

We know how to craft effective legislation and ensure its passage. The American Clean Energy and Security Act of 2009 failed because at just over 1,200 pages it was too short. The 2,700 page Patient Protection and Affordable Care Act showed the way "Forward." By creating lengthy bills and rushing them to a vote, we can defeat the enemies of fairness, real science, and Mother Earth.

A-Not-Too-Bright Future: There is reason for optimism. As global temperatures rise, so will the dollar value of government contracts and grants. Together, we can warm the hearts of progressives everywhere by ushering in an industrial ice age.

[End of excerpts from GAIA brochure.]

We could go on and on about our salvific projects, but you get the idea—we're sincere, creative, and above all committed. Pending sufficient contributions, we hope to have the brochure ready for the next fabulous and momentous UN climate change conference.

SideBar25: Earth Day 2015: Thought for the Day

To paraphrase Archie Bunker, "Earth Day is once again at our throats."

Well, here's a wishful thought for the 45th Earth Day: Let's take all the hyperventilated resources wasted since the early 1980s to fuel public panic over supposed manmade global warming and use that energy to power the planet into the next century—*what a wonderful world this would be* (H/t Sam Cooke).

All the talk about saving the earth should have been focused on action to lift a billion people out of poverty. But, as the saying goes, talk is cheap. And, caring for the planet's poor is expensive, and hard, especially in terms of getting the world's ruling class to even budge on their arrogant leftist ideology that aids in the impoverishment of the masses.

Grandiose schemes designed to rescue Mother Earth never seem to include "her children." Whether from the 15th United Nations Climate Change Conference in cold Copenhagen in 2009, or from unseasonably chilly Cancun in 2010, or Durban in 2011, Doha 2012, Warsaw 2013, Lima 2014, or the 2015 Paris affair, it's easy from the opulence of endless climate-change summits to demand ecological salvation. The hard part is saving anything except maybe selfie self-esteem and bureaucratic booty.

Ambassadors for Global Warming

SADLY, IN JULY 2013, a group of about 200 evangelical scientists and academics sent a letter to House Speaker John Boehner, Senate Majority Leader Harry Reid, and Members of Congress advising them to take immediate action "to reduce carbon emissions." The letter read in part, "As evangelical scientists and academics, we understand climate change is real and action is urgently needed."

The letter goes on to preach the tired environmental-activist mantra listing the results of man's sinful actions like hottest year on record (for less than two-percent of the Earth), wildfires, droughts, and public health outbreaks (all unfortunate but not too atypical).

The sin of course is modern human activity in prosperous nations. You know, the kind of sinful activity that has lifted so many out of poverty, protected them from never-ending "unusual" weather events, and dramatically contributed to healthy living. The very same kind of activity made possible by God's grace.

After all, this activity involves the use of God-given, abundant, inexpensive natural resources—the kind that can truly benefit those Jesus called "the least of these."

But, instead of advocating for good stewardship—which is reasonable and right—and wide, wise distribution of what is readily available to those in dire need, hundreds of evangelicals (mostly biologists) armed with a superficial understanding of atmospheric science and a misapplication of scripture chose to advocate for some sort of climate justice.

As a Christian and an atmospheric scientist with 35 years of experience, I support the efforts of my fellow Christians for considerate use of God's provisions. However, I have frequently witnessed the metamorphosis, refinement, and exploitation of the never-ending story that humans are somehow destroying a "fragile" planet—this time by releasing too much carbon dioxide.

Of all people, Christians should be the least gullible on this untenable position, since Christians have a solid foundation build on the belief that God is creator and sustainer of all things. Forget about the dubious "sin" of anthropogenic global warming; Christians should focus instead on the arrogance germane to the idea that not only are humans causing long-term global climate change, but that we can fix it.

A recent scholarly book has looked into the evolution of modern Christian thought on the global warming issue. *Between God & Green: How Evangelicals Are Cultivating a Middle Ground on Climate Change* by Katharine K. Wilkinson (Oxford University Press, 2012) states that "[c]limate change…is a critical piece of the new evangelical politics emerging in American public life." And so it is… unfortunately.

Between God & Green explores the concept of "creation care," or more specifically, "climate care." The book primarily follows the conversion and progress of creation care, evangelical elites who advocate for climate change politics, policy, and personal commitment. *Between God & Green* is fairly thorough in its in-depth exploration of the world of evangelical thought on human's relationship to the environment. The book's assessment of the role of Christian eschatological (end time) beliefs on how a person views and treats the environment is particularly revealing.

Between God & Green, however, is somewhat slanted. It portrays the climate care movement's leaders and fellow believers as more-or-less apolitical, moderate, and Spirit-led, while those who challenge their faith—that mankind is responsible for the sin of climate change—are painted as essentially right-wing dupes of the Republican party. In chapter 5, "Engaging People in the Pews," the author summarizes interviews she has had with congregants regarding a position statement of the Evangelical Climate Initiative (an organization whose funding "comes largely from secular foundations that support conventional environmental advocacy"). ECI's position statement, which is reproduced in the book's appendices along with eight other creation-care organization documents, is titled "Climate Change: An Evangelical Call to Action." The author observes that "conservative politics and conservative Christianity and the influence of the echo chamber of conservative media were evident in group discussions, as churchgoers echoed these discourses." Ignored, as always, is the biggest echo chamber of all, academia.

But, that ignorance aside, the attempt to relate climate care activism to a scriptural foundation throughout *Between God & Green*, gets to be a bit over the top at the book's conclusion. The book ends with a curious application of the events in the gospel of Mark, chapter 11. Here in the gospel is where the money changers and those selling doves for sacrificial offerings are expelled by Jesus from the temple in Jerusalem. The author interprets the event this way, "Jesus challenged what had become business as usual—politicization and commercialization of this sacred space, which benefited the privileged and preyed on the poor...." The author then went on to say that "[s]imilarly, climate care challenges entrenched perspectives and practices perceived to be economically, morally, politically, and theologically corrupt, with the hope of installing more authentic ones in their place." Alternatively, when reading the scriptural passage, I almost immediately thought of carbon trading and selling—a scheme that will further impoverish the middle class, enrich unscrupulous investors, and, as usual, do next to nothing to help the poor, who always seem to miss out on these massive wealth transfer schemes. Now that is truly economically, morally, politically, and theologically corrupt.

The climate of the world will be much better off when Christians focus on being ambassadors for Christ, rather than activists for the atmosphere.

SideBar26a: Go Tell It On the Mountain

As the word of the U.N. Intergovernmental Panel on Climate Change (IPCC) report goes forth *ex cathedra* to make disciples of all nations, many will embrace it, some will even adore it. Verses from the AR5 (Assessment Report #5) version of the IPCC climate bible will be quoted *ad nauseam* to win converts and to support sanctimonious government programs designed to limit the liberation of the masses.

Sadly, of all believers, it will be gullible Christians who arguably will be most culpable for the ensuing continued misery.

It's been estimated that more than a billion people still live in poverty. Access to low-cost, abundant, readily-available, God-given natural resources would go a long way to alleviate this real-world crisis. Yet, rather than having faith that God will sustain His environment so that the liberating word of Christ can go forth, Christians have put their trust in the U.N.'s "arm of the flesh."

The IPCC has been preaching for decades that human souls are guilty of raising temperatures worldwide. Yet the IPCC's prophecy has not materialized. Why not? Because the high priests of climate science have too little faith. They trust in carbon dioxide, which comprises only 0.04% of the atmosphere, to perform miracles.

As argued throughout this book, the reason why the global temperature trend has been nearly level for more than 15 years now as paltry carbon dioxide increased is quite likely explainable by water's role in climate control. It seems likely that God wisely assigned the role of climate regulator to water in all its phases and characteristics—water in the invisible vapor form, liquid form (oceans, rainfall, clouds), and ice form (glaciers, snow, clouds); water transport and distribution across the globe; and, the energy of conversion associated with water's phase changes. Because of water's immense complexity, venerated climate models do a poor job properly simulating water's role in long-range global climate reality. Yet so many of the faithful continue to trust in the power of man-made "carbon pollution" and continue to fret about "climate justice" nonsense. But, even a papal encyclical cannot make disastrous manmade climate change palpable to reality.

Advice to Christians: Go tell it on the mountain. Preach the Word, both in season and out of season, for:

> *While the earth remains,*
> *Seedtime and harvest,*
> *And cold and heat,*
> *And summer and winter,*
> *And day and night*
> *Shall not cease. [Gen. 8:22, New American Standard*
> *Bible (NASB)]*

Now, there's a long-term, global climate forecast you can really trust.

SideBar26b: When the Saints Go Marching Out

Thousands of true believers were in New York City on Sunday, September 21, 2014 for the People's Climate March to profess their faith in the power of greenhouse gases. These trace chemicals (especially carbon dioxide) have been imbued with such awesome characteristics as to draw the faithful from the ends of the nation and beyond.

Left-wing environmentalist dogma demands that 0.04% of

carbon dioxide in the atmosphere will bring about hell on earth—
overwhelming the variability of water vapor at 0% through 4%
concentration and the liquid and solid phases of water (besides the
energy associated with the phase changes of water, and oceanic
circulations).

Given the stabilizing qualities of water, it's not surprising that
real-world data have belied the prophesies of doom. Yet Mother
Earth's congregants soldier on. Such is blind faith.

The faithful still dream that tragedy will strike or at least that
they can continue to use the specter of impending doom to
proselytize others.

This is all well and good for so many in today's world who
have nothing else to cling to other than the hope that humans are
causing long-term, global climate change. They now have a cause
loftier than themselves to champion.

But, what about marching Christians who sincerely hold to
the Bible as the Scripture of Truth? Well, maybe the shame the
President is trying to lay on countries who do not believe in the
sinfulness of carbon dioxide emissions, should be turned on those
who hold the Word so lightly.

After all, doesn't the Bible assure that "While the earth
remains, Seedtime and harvest, And cold and heat, And summer
and winter, And day and night Shall not cease" [Gen. 8:22, NASB]
(See previous sidebar).

And, isn't the same Bible from Genesis to Revelation
thoroughly focused on *people*?

One Christian organization relying on the reality of
atmospheric dynamics while defending the world's poor from airy
misanthropic ideologies is The Cornwall Alliance for the
Stewardship of Creation. (Note that *Between God & Green* implies
that the Cornwall Alliance is somehow anti-environment. To the
contrary, the Alliance's mission focuses on real aid to the world's
poor <u>and</u> does not neglect effective care for creation.)

The Alliance released "Ten Reasons to Oppose Harmful

Climate Change Policies" [ix]

This declaration makes reasonable assertions from a Christian perspective including stating that "As the product of infinitely wise design, omnipotent creation, and faithful sustaining (Genesis 1:1–31; 8:21–22), Earth is robust, resilient, self-regulating, and self-correcting." And, "Earth's temperature naturally warms and cools cyclically throughout time, and warmer periods are typically more conducive to human thriving than colder periods."

The Alliance further calls on Christians "to practice creation stewardship out of love for God and love for our neighbors—especially the poor."

So, instead of saints marching out to a dirge that is likely to prolong and maybe even expand poverty, perhaps they should come marching in to the hopeful tune of the gospel that is truly good news for people and the planet.

[ix]   www.cornwallalliance.org/2014/09/17/protect-the-poor-ten-reasons-to-oppose-harmful-climate-change-policies

Cap-and-Trade Liberty

MODERN HISTORY OFFERS a simple lesson: it's far easier to enact progressive laws and programs than it is to stop them once enacted. Bills that fail to pass can be rewritten and submitted to future Congressional sessions. Broad programs that can't be implemented in one fell swoop may be accomplished incrementally. Progressives only have to succeed once to get their way. To prevent progressives from succeeding, opponents must successfully block every attempt.

Though the American Clean Energy and Security Act failed to pass in 2009, the idea of a national cap-and-trade program is far from dead. In fact, it's still very much alive—though flying under the radar. For example, in late 2011 California passed its own EPA-inspired greenhouse gas cap-and-trade program, the Global Warming Solutions Act. Will the people realize that these programs make their lives more difficult in time to stop them?

If the Patient Protection and Affordable Care Act were fully implemented, the government would control your body. And with caps on carbon emissions, the government would control much of what's left: how you keep warm (more wind mills, less natural gas, and no coal); how you have fun (no big, fast cars or boats and fewer long trips); and what you eat (less beef, less pork, more soy). Products and activities that green zealots deem wasteful or ostentatious will be targeted for control. As a result, products and activities that were once within the reach of most people will only be available to the allowance buying/trading elite.

We are told there is a consensus among scientists that disaster is sure to follow if we continue in our errant, carbonaceous ways. Is there also a consensus among scientists that greater government control over our lives is the correct solution? There seems to be. To treat one interpretation of scientific evidence as incontestable merely because it is the consensus interpretation is to abandon skeptical inquiry in favor of dogma. To link that dogma to a specific public policy prescription is the height of arrogance. If this trend continues, then the people will be disenfranchised—the power to govern will rest solely in the hands of elite specialists.

Contrary to what you may have heard, atmospheric scientists who question the reigning dogma that humans are the cause of long term global climate change are *not* against energy efficiency, conservation, common sense rules and regulations for industry, and reasonable guidelines for the use of raw materials. Like most American citizens, we generally oppose government control over our lives and livelihoods, and we are particularly suspicious when such control is based on speculation about what might happen in the future. And we are doubly suspicious when we are told that it's not based on speculation—it's based upon knowledge of the future that is very close to absolute certainty. Because as practicing scientists we understand that data is open to different interpretations; that the number or credentials of scientists who believe something has no bearing on whether it is true; and that when scientists resort to name calling and bullying something has gone very wrong. Science flourishes when scientists disagree passionately yet respectfully. When the majority insists that certain ideas and theories are "good science," that only those ideas and theories should be taught in schools, and that only those ideas and theories should inform public policy, then it's no longer science. It's dogma and it can only lead to authoritarianism.

Good science can withstand questioning, criticism, and doubt.

In fact, good science demands it. Science is not advanced in the same way that a building is constructed. It's not just a matter of adding bricks to an unshakeable foundation. At times, scientific progress requires a paradigm shift. Science historians and philosophers of science may have different views about exactly how this occurs, but it's clear that science is much more tentative than some scientists admit. In the early 20th century, physicists discovered that subatomic particles don't obey Newtonian physics. It took at least twenty years to develop a new theory (quantum physics), and there were plenty of skeptics—Albert Einstein was one of them. The theory survived but there are many unanswered questions.

The situation in contemporary climate science is very different. Instead of admitting that there are still many unanswered questions, we are told to trust experts who purportedly possess near-perfect knowledge about today's climate and the ability to divine future atmospheric conditions. Never mind that short-term climate predictions (not just weather predictions) have been blatantly wrong. Never mind that the climate system is highly complex and only partially understood. And never mind that there is no track record for long-term climate forecasts. Because if you don't trust the sanctioned forecasts about what the Earth's climate will look like in 50 or 100 years, then according to the experts you aren't a real scientist.

One prediction that can be made with a high degree of confidence is that the imposition of carbon rules will be a major setback for individual liberty. The more restrictions there are on what private companies and individuals can do, the fewer choices will be available to individuals. When government obtrusion leaves individuals with fewer choices, it increases the average person's level of servitude. When government rules and regulations determine how we stay warm, what type of vehicle we drive, what food we eat, and even what type of light bulbs we can use, then we have achieved what could be fairly characterized as a cap-and-trade on freedom.

In addition to this expected downside, there will be an "upside" in the form of soaring energy prices. As President Obama freely admitted, "under my plan of a cap-and-trade system, electricity rates would necessarily skyrocket." Does he understand that this also means higher food prices, clothing costs, and travel expenses? That won't just hurt the poor—it will swell their ranks.

The President did promise change. He did promise greater emphasis on science. However, it's a very selective form of science for selective purposes. Restricting carbon emissions won't ensure that the US leads the world in scientific developments. But it might ensure the opposite. Most of the progress in science comes from advanced, industrialized countries—not countries with crash programs built around 19th century wind turbines.[74]

The health of our planet and its inhabitants requires a sound, science-based energy program. Such a program would focus on developing more clean and safe energy—not less. Such a program would be unbiased; it would look at ways to make the use of fossil fuels and nuclear energy cleaner and safer as well as develop so-called "renewable" energy sources.

The elite are perfectly content to see gasoline hit $5 per gallon. They can afford to pay extra, while rest of us will have to do without.

For the health of our planet and especially its inhabitants, any cap-and-trade carbon reduction system must be capped-and-exchanged for a more sound science-based energy program. And if a reasonable, compassionate, free people have their say, it will be.

One Step Forward, Two Steps Back

The battle over the government's right to cap or restrict carbon emissions in response to a perceived threat continues. On June

20, 2011, the U.S. Supreme Court issued a ruling on *American Electric Power (AEP) v. Connecticut*, a case ostensibly about whether states can demand that utilities located outside their borders reduce their greenhouse gas emissions.[75] In a decision that was both good news and bad news, the court ruled 8 to 0 in favor of AEP. On the surface, the ruling was a major blow against climate tort cases. However, since the ruling reaffirmed the high court's 2007 decision in *Massachusetts v. Environmental Protection Agency*, which said the EPA has the right to regulate greenhouse gases through the Clean Air Act, the case mainly served to confirm that the federal government can limit citizens' choices based on a mere hypothesis about the long-term global effects of using fossil fuels.

There is probably no one better qualified to expose such regulations for what they really are—an assault against individual liberty—than the veteran opponent of climate-science-by-consensus Patrick J. Michaels[76] and his colleagues specializing in health, economics, education, law, national defense, international development, and other fields.

In the book Climate Coup: Global Warming's Invasion of Our Government and Our Lives,[77] Dr. Michaels explains:

> *When students are threatened with death from [the consequences of] global warming, when our military raises the threat of war from global warming, when the state has the apparatus to run our lives because of global warming without any additional legislation, when our Congress legislates tariffs that could provoke trade wars because of global warming, when the threats of global warming to the developing world are egregiously exaggerated, when the biomedical community hypes unfounded health and mortality fears, and when the scientific peer-review process becomes skewed against anything moderate, we have witnessed a coup. Global warming has taken over our government and our lives. (pp. 12-13).*

In other words: Lives are being threatened, our national security is being put at risk, we are being stripped of our individual rights, and science is being corrupted. All of this is owed to facts that cannot be independently verified—"facts" pulled out of proxy data, computer models, and predictions about what might occur 50-100 years in the future. *Climate Coup* offers plenty of challenges to the dubious science and detestable academic tactics used to promote the theory that ordinary human activity, if not stopped, will destroy the planet. However, it's the discussion of political and economic implications that stands out most.

In Chapter 1, "The Executive State Tackles Global Warming," veteran legal scholar Roger Pilon and attorney Evan Turgeon conclude:

> *The Progressive Era [in the early 20th century], enamored of science, sought to bring about rule by 'experts' ensconced in government planning bureaus. We have that pretty much today. As a result, environmental decisions involving not simply science but, at bottom, value-laden tradeoffs are made by relatively few unaccountable bureaucrats concentrated in the executive branch, with only sporadic and uneven judicial and congressional review. (p. 41).*

Progressives often claim to be champions of science. But they conveniently forget that science is based on facts and logic. When they cite the opinions of experts, they are committing one of the most common logical fallacies—they are making their case based on an *appeal to authority*. By what standard are these persons deemed "experts"? And even if they are universally acknowledged to be experts, that doesn't mean they are always right. Besides, it's not unusual to find different "experts" on opposite sides of the same issue. Real science is based on iron-clad logic—not logical fallacies.

Anthony J. Sadar

In Chapter 5, "Climate Change and Trade," agricultural economist Sallie James observes:

> Politicians have made almost an art form of trying to defy economic gravity. The climate change debate has provided numerous examples of their attempts to deny the existence of tradeoffs and unintended consequences. They want to be seen to be doing 'the right thing' by the environment, and to please certain special-interest groups. But the policies they have proposed to combat climate change negatively affect other special-interest groups, whose votes and campaign contributions are at least equally valued. (p. 156).

Similarly, the theory that human activity will lead to a global catastrophe by the end of the century denies the existence of major, non-human factors affecting the Earth's climate. Or if it does recognize that there are other factors affecting the Earth's climate, it assumes that they will become dormant. Anthropogenic global warming alarmists are like business planners who assume that competitors will go into hibernation while they prepare and execute a new strategy.

The authors of *Climate Coup* use facts and figures to show that even if human greenhouse gas emissions are causing climate change, the cure is likely to be worse than the disease. It doesn't make sense to take one step forward and two steps back. It makes even less sense when you know that hard-earned liberties will be trampled.

SideBar27: The Global Warming Hypothesis Meets Reality

As discussed earlier in this book, there is so much bluster these days from the White House and one of its political enforcers, the EPA, on the "certainty" of human-caused climate change and the urgency of saving the earth from prosperous people.

But, like the climate itself, the hypothesis of man-made climate change cycles through history.

Take recent history for example. In the 1970s, the hypothesis was that the globe was potentially headed for the next ice age. I know this not only because of pronouncements from popular press at the time, but also because, as an undergraduate student at one of the top schools of meteorology, Penn State University, the buzz was that the global climate was moving toward seriously colder conditions. To substantiate this claim, professors referenced not only recent climate trends and observations but also the work of respected scientists, such as astrophysicist Milutin Milankovitch, who had investigated long-term climate cycles. According to the "Milankovitch Theory," which is based on cyclical variations of the Earth's orbit around the sun, the globe was heading for some big-time icy changes.

Even outside the college campus, the culture was primed for the next ice age as evidenced by a Christian tract by Walter Lang and Vic Lockman. The pamphlet asked in its title *Need We Fear Another Ice Age?* And, Leonard Nimoy went *In Search of the Coming Ice Age* on TV.

Of course, technological fixes were proposed like adding coal dust to the surface of advancing glaciers so that the encroaching ice would absorb more solar energy and melt away.

Yet, what a difference a few decades make.

Today, it is fashionable to expect disaster from too much warmth. So, the smart money is on promoting dire predictions and consequences of rising thermometers, even in the face of no global warming for more than fifteen years.

From my many years of experience in the atmospheric science profession as an air-pollution meteorologist, air quality program administrator, and science educator, I can attest the fact that long-range, global climate-change outlooks are nothing but insular professional opinion. Such opinion is not worthy of the investment of billions of dollars to avoid the supposed catastrophic consequences of abundant, inexpensive fossil fuels and, subsequently, to impoverish U.S. citizens with skyrocket energy costs.

I have conducted or overseen a hundred air-quality studies, many using sophisticated atmospheric modeling. Such modeling—comparable to or even involving the same models as those used in climate modeling—produced results for relatively short-term, local areas that, although helpful to understanding air quality impact issues, were far from being able to bet billions of taxpayer dollars on. Yet, similar climate models that imagine conditions for the entire globe for decades into the future are used to do just that—bet billions of taxpayer dollars.

Bottom line, nobody can detail with any billion-dollar-spending degree of confidence what the global climate will be like decades from now. But, it's easy to predict that given enough monetary incentive and the chance to be at the pinnacle of popularity some climate prognosticators—and certainly every capitalizing politician—will continue to proffer convincing climate claims to an unwary public.

CSI: Climate Science Investigation

"IF EVER CIRCUMSTANTIAL evidence pointed to a criminal it does so here," Dr. Watson remarked.

"Circumstantial evidence is a very tricky thing," answered Holmes thoughtfully. "It may seem to point very straight to one thing, but if you shift your own point of view a little, you may find it pointing in an equally uncompromising manner to something entirely different." [78]

In the wake of Climategate, hockey stick curves, and wayward temperature trends, the sferics[79] surrounding global-warming continue to charge the atmosphere. The drama is reminiscent of made-for-TV murder mysteries. In crime scene investigation shows, the audience is presented with not only physical evidence but interviews with detectives, family members, witnesses, and even the defendants.

Later, when the convicted person is interviewed, they often say something such as, "The cops suspected me from the very beginning. They never even looked at anyone else. Anything they found that pointed to me as the guilty party, they took seriously; anything that pointed to someone else, they dismissed."

When this kind of statement is made by someone sitting behind bars, the viewer is understandably suspicious. After all, a person who has been convicted may still hope to win on appeal and be released. A person in that position might say anything that they think will help.

Then again, what if what they are saying *is* true?

Whether a claim is true or not comes down to three things:

- The evidence itself;
- How the evidence is interpreted and used; and
- The credibility and believability of the person interpreting the evidence.

However, something could go wrong at any point in the process. For example, assume the person interpreting the evidence has great credentials and is highly-skilled, experienced, and—to top it off—very popular. This person may be seen as someone who never makes mistakes; therefore, his or her word is sacrosanct.

The evidence is often overshadowed by the interpretation.

The analogy to a climate science investigation should be obvious. All scientists are confronted with the same data. But just as in a crime scene investigation, what matters most is how the data is interpreted and the credibility of the person or persons doing the interpreting.

For example, temperature-based evidence of climate change comes in multiple forms: direct thermometer readings, direct satellite sensing, and historical proxy data (indirect indicators such as written accounts, analysis of tree rings, and ice core measurements).

Satellites provide accurate temperature readings for the troposphere layer (roughly the lowest 7 miles of the atmosphere in the mid-latitudes) going back to 1979, when orbiting thermal probes first became operational. Thermometers located throughout substantial portions of the Earth's land mass provide accurate readings going back to about 1850. There is proxy data that can be used to deduce temperatures going back thousands of years and even hundreds of thousands of years. However, the accuracy of such data depends on the type of proxy—i.e., the source and how the data is derived from the source. Written accounts and tree rings only go back a few thousand years; ice cores go back hundreds of thousands of years.

David Keeling in Mauna Loa, Hawaii began monitoring atmospheric carbon dioxide (CO_2) concentrations in the late 1950s. The measurements indicate a sharp rise in global CO_2 values—from approximately 310 ppm at the start to about 400 ppm today. Scientists have wondered how this increase, apparently from anthropogenic carbon emissions, has affected the earth's atmosphere, particularly since the rise in CO_2 levels seems to track the rise in average worldwide temperatures. (Though in recent years, as CO_2 concentrations continue to rise, global temperatures have been leveling off.)

Historically, the focus has been on CO_2 as the culprit and increased terrestrial temperatures as its crime.

A star lawyer for the global warming plaintiffs is former Vice-President Al Gore, who argues that "the earth has a fever and...maybe that's a warning of something seriously wrong." A star witness is James Hansen, director of NASA's Goddard Institute for Space Studies, the expert behind many of Mr. Gore's claims. Coal-fired power plants, in particular, have been fingered for arrest by Dr. Hansen.

No one doubts the sincerity of former Vice President Gore and Dr. Hansen. However, sincerity (like good intentions) is not a virtue. It doesn't guarantee that someone is correct—or even on the right side. Politicians and even scientists can be sincerely wrong. Unfortunately, in the courtroom of public opinion both gentlemen are extraordinarily convincing and believable—proving once again that the evidence may be overshadowed by the interpretation and the believability of those doing the interpreting.

A growing number of independent earth and climate scientists suspect that the wrong perpetrator has been apprehended. Only when the evidence is examined dispassionately with the assumption that the perpetrator has not yet been identified, considering all plausible suspects (including cosmic rays and sunspots), while treating them all as innocent until proven guilty, are we likely to discover the truth about climate change.

Meanwhile, humans have already been convicted by the media. But this is why we have judges, juries, and due process of law. The media's mission is to maximize its audience. The judicial system's mission is to see that justice is served. Historically, the court of science has always respected those who raise reasonable doubts. But when there is tampering with evidence, a packed jury, and slander against the scientists who raise reasonable doubts, then you know you are dealing with a kangaroo court. And the usual suspect—manmade greenhouse gas—may indeed be innocent of crimes against the environment.

SideBar28: Climate Science in *Jeopardy*

Scientific practice is a bit off these days. Like the popular game show *Jeopardy*, it seems that the desired response is given to the "correct" question that elicits that response. But, the answer-in-the-form-of-a-question is always the same.

Responses seem endless, let alone contradictory:

The reason why thermometers are rising so quickly worldwide; or, the reason worldwide temperatures have leveled off in the past 18 years.

The cause of the higher-than-average hurricane season in 2005; or, the cause of the lower-than-average hurricane season in 2013.

The reason there has been so little snowfall in the U.S. and Europe; or, the reason there has been so much snowfall in the U.S. and Europe.

And on it goes.

The answer is always: "What is man-made climate change?"

Not every "unusual" weather event evokes the patent humans-are-responsible response, however. Oftentimes, to attract unwary audiences, not so unusual, but unfamiliar, events are exaggerated by purveyors of pernicious prognostications.

Take the "polar vortex" scare. This natural phenomenon was proffered as something new, something frightening, something produced by people living comfortably. Of course it is none of that. This is verified by the *Glossary of Meteorology* published by the American Meteorological Society in 1959, where this well-known phenomenon was clearly defined, not hyped.

As the ancient Ecclesiastes writer observed, there really is "nothing new under the sun."

Certainly the Intergovernmental Panel on Climate Change (IPCC) practices the sort of science that gives the answer first, then seeks the appropriate corresponding questions.

The IPCC defines its role as "to assess...the risk of human-induced climate change, its potential impacts and options for adaptation and mitigation." In other words, the IPCC assumes from the get-go the answer that anthropogenic climate change is a fact. It is then the game of researchers, enticed with prized government grants, to find the questions that always lead to that answer.

The U.S. Environmental Protection Agency (EPA) also gets in on the act. The *Federal Register* recently announced that the EPA is looking for research participants for their new campaign to fight the impact of climate change (ostensibly human-caused) on public health. Healthy skeptics need not audition.

This is not how science is supposed to work. It's not supposed to be like some rigged game show with fabulous prizes for coached contestants, and self-serving promoters including environmental and social activists, career politicians, bureaucrats and technocrats, crony capitalists, *incite*ful journalists, bloggers and PR spinners.

Science must ask the questions first, then work diligently to ascertain the right answers. In fact, the true responses to climate questions such as: "Why has the globe not shown a substantial temperature rise for more than 18 years; and, why has there not been an increase in frequency of U.S. hurricanes?" are still being fervently researched. As they should be.

So, to keep climate science out of jeopardy, and to change the channel on this long-running popular program that sponsors foregone conclusions, fruitful scientific answers must be produced by fertile questions.

After all, the *honest* correct answer is: "The climate system is much too complex for hasty arrogant assertions of long-range global conditions."

Climate Science Deserves Diversity, Too

DIVERSITY IS ALL the rage on college campuses. Unfortunately, it's racial, ethnic and sexual diversity that are being promoted—not diversity of ideas and opinions. In most fields, and particularly in complex fields such as climatology, diversity of thought should be encouraged to help sort out the problems, causes, and solutions. After all, no one person or group has a monopoly on the truth.

A careful and honest study of the earth-atmosphere system leads to one obvious conclusion: we know very little. Mountains of data have been gathered from the Earth's land, oceans, and atmosphere. There are some things we can deduce from the data with just a little analysis. For example, we've learned about the atmosphere's structure, chemical composition, and variability (both short-term and long-term). However, our ability to integrate and interpret the data for the purpose of making long-range projections is much more primitive. And yet many scientists and politicians speak with one voice, claiming that we know with confidence and precision how the climate operates and how its operation may be safely altered.

Where did such single-mindedness originate? The halls of academia are a good place to look. Atmospheric science has blossomed since I was a meteorology student at Penn State University in the mid-1970s. Back in those days, the field was dominated by slide rules, paper maps, and batch computing. Since then, the field has exploded, not only with climate models running on lightning-fast computers, but in celebrity as well. Meteorologists have handled the increased attention and authority in a reasonable, measured manner. To this day, it's still their job to churn out daily and weekly weather forecasts for Podunksville, USA. Climatologists, however, were not accustomed to being in the limelight, and seemed to relish what was at times fawning attention. Climatology was suddenly promoted from a cloistered profession (mainly engaged in the tedious work of compiling facts and figures) to a field entrusted to issue definitive diagnoses and prognoses about the health of planet Earth extending decades into the future. While Meteorologists can tell you whether to take an umbrella, Climatologists, it is now imagined, can tell us whether the Earth will still be inhabitable in 50 years and which energy sources we should invest in.

Climatologic reports and presentations, with their marvelous mathematics and three-dimensional animated graphics, meet or exceed technical expectations. To many people, complex calculations and flashy graphics signify knowledge and certainty. However, by making predictions about the distant future, climatologists are behaving like carnival fortune tellers. It's great entertainment and business is brisk. They've discovered what keeps people coming back for more: tales of narrow escape. The Earth is doomed, but if we go back to using oxcarts and compost piles, we just might survive.

Back at the academy, as climate models continued to impress, education models digressed.

Climate science needs more diverse insights and viewpoints to help make sense of mountains of data. More diverse voices are needed to help the public see through the politicization of science. Most important, different viewpoints are needed to counter outrageous claims—such as the claim that all "real scientists" agree that humans are dramatically altering the Earth's climate.

Actually, diverse voices are speaking out, but their reach is limited by media channels that abridge, filter, or completely ignore perspectives that challenge the science establishment. One reason for this lack of coverage is that many journalists and editors don't understand how great scientific discoveries are made. They don't understand that most of science is never really "settled." For example, Isaac Newton's view of the physical universe was amended by discoveries in electromagnetism, relativity, and quantum mechanics. Science progresses most when scientists are free to propose, test, and debate different hypotheses and theories.

Science is in peril when challenges to popular theories are quashed for political reasons. But that's precisely what appears to be happening in climate science. Conversely, research serving specific political purposes (e.g., the infamous hockey stick graph) is immediately hailed, declared conclusive, and rushed to the public. Meanwhile, events and practices that should raise red flags are neatly papered over, such as the Climategate scandal, the IPCC's stacked deck approach to research and reporting, and doubts about the representativeness of temperature measurements in North America and elsewhere.

These recent ominous signs cast doubt on the integrity of contemporary climate science. The proper response should be to call for more diversity of thought and more overt reasonable dissent. Good science isn't sheltered—it's battle-hardened. At a minimum, climate scientists should be more diligent, open, and honest about what they know and, more importantly, what they don't know.

Weird Science

TODAY'S SCIENCE IS getting weird. On one hand, anyone who is unconvinced by the evidence and consequently challenges climate science orthodoxy is labeled "anti-science." On the other hand, respectful attention is paid to avowed Luddites such as author Kirkpatrick Sale, who applauded the Unabomber as the "last person to do any [thinking]" regarding the negative consequences of modern technology.[80] In other words, if you suggest that higher standards must be met before popular conclusions can be pronounced certain, then you are an enemy of science. But if you voice the concern that "high tech has such a vast environmental impact...that society will no longer be able to cope," as did Sale, then you are a friend of science.

Even as the evidence continues to soften, anthropogenic global warming (AGW) picked up another celebrity endorsement in early 2010. Osama Bin Laden reportedly expressed disappointment that more was not accomplished at the December 2009 Copenhagen climate summit. He even joined the environmentalist chorus identifying the US and other large industrialized nations as the world's evil actors. Poking his head out of his cave just before Groundhog Day, bin Laden did not see six more weeks of winter, but he did see continued global warming if Western countries do not scale back their industrial activity. AGW supporters hardly flinched when the world's most notorious terrorist joined entertainers, demagogues, and Luddites in bemoaning climate change. Instead, they lamented that "Drudge, Fox News and other right-wing media seize on al-Qaida leader's taped comments..." [81]

While AGW proponents consider Osama bin Laden's endorsement unwelcome, they don't appear to have a problem with endorsements from elitist Hollywood celebrities who lack scientific credentials. Given the disputes and unenforceable agreements from the Copenhagen climate summit and subsequent climate change gatherings, radical environmentalists have redoubled their efforts to make anthropogenic global warming the only view of climate change accepted as scientific. They'll take almost all of the help they can get

For example, the National Center for Science Education (NCSE), an organization established for the express purpose of promoting Darwin's theory of evolution in schools, has expanded its mission to "keep out [of classrooms]... climate change denial." The NCSE defends the teaching of science by calling a piece of legislation that merely encourages (i.e., permits) teachers to present the scientific strengths and weaknesses of controversial theories the "monkey bill." [82]

On the international stage, an organization called the InterAcademy Panel on International Issues (IAP) is stepping up efforts to promote progressive science. The IAP, founded in 1993 in New Delhi, is a global network of more than 100 national science academies—representing countries from Afghanistan to Zimbabwe. The IAP states that its primary goal is "to help member academies work together to advise citizens and public officials on the scientific aspects of critical global issues." The IAP publishes statements on issues including population growth, human reproductive cloning, science education, science and the media, biosecurity, evolution, ocean acidification, and (naturally) climate change. [83]

Insight from a variety of perspectives is valuable to science. For the IAP to speak as the world's ultimate authority on science when it only represents its member academies' point of view, though, ignores the legitimate views of non-member academies and societies, industry scientists and engineers, and even private consultants, each of whom may have valuable insights about how to address global challenges. Just as the NCSE hopes to guide national policies in the US, the IAP strives to gain undue influence over international policies to advance what it considers rational facts (that AGW is certain) and to oppose what it considers irrational opinion (that there are legitimate reasons to doubt AGW).

Since most policymakers are not climate scientists, it's understandable and beneficial that they rely on people with credentials in the field and input from a variety of relevant interest groups. Still, even people who are not trained in atmospheric science can use common sense and a healthy dose of skepticism to evaluate what they are being told. For example, you don't need a PhD in the sciences to recognize that forecasts made in the early 21st century about what the world's climate will look like in the year 2100 should not be viewed as infallible or even probable.

Any group that believes it is vastly superior to the *hoi polloi*, and has the arrogance to say so, deserves to be treated with suspicion. Yet this is precisely the attitude of many groups that promote AGW. Climate science, along with most other physical sciences, has strayed far from its traditional stance. At one time, humility was encouraged and treasured. Now, it seems to be undesirable—a sign of weakness. Concomitantly, today's scientists seem to have few qualms about flaunting their "expert" status, demanding government largesse, and ridiculing leaders and public policies with which they disagree.

Scientists breaking ranks on AGW are predominantly atmospheric scientists and professors (such as aerospace engineer Dr. Willie Soon of the Harvard-Smithsonian Center for Astrophysics and meteorology professor Dr. Richard Lindzen of MIT) who have either retired, do not rely on government grants for their research, or are just extraordinarily gutsy.

Sadly, here is what climate science looks like today: If you're promoting AGW from the hallowed halls of academia, a science-body cathedral, or even a clandestine cave, then you're advancing science. If you're raising doubts about AGW, whether from a top-flight university or an exemplary private enterprise, then you must be either a right wing extremist, a paid operative, or mentally ill.

SideBar30a: When You Can't Think, You *Need* to Think

Sometimes TV commercials can be downright inspirational (h/t DirectTV)...

When your climate change theory's on the fritz, you get tense.

When you get tense, you can't think.

When you can't think, you *need* to think.

When you need to think, you start to worry.

When you start to worry, you think your theory has to survive.

When you think your theory has to survive, you start thinking wild thoughts.

And, when you start thinking wild thoughts, you chase imaginary theories into something highly unlikely (like "Human carbon pollution is a weapon of mass destruction").

Don't chase imaginary theories into something highly unlikely.

Get rid of imaginary theories, and upgrade to direct observations... and what they say about the real world of climate change.

Anthony J. Sadar

SideBar30b: Help for Sufferers of CD (Climate Dysfunction)

With the release this past Monday of new Environmental Protection Agency directives to cripple coal-fired power plants, the effort continues to rid the U.S. of the disease of low-cost energy.

A big new concern being pandered by the EPA is the effect of coal-induced climate change on public health. But, what if fossil fuels aren't as big a burden on long-term global climate change as so many that mix politics, ideology, and science believe? And, what if, as real-world data is proving, the predictions of such "climate disruption" is more fantasy than fact? The real issue then becomes a mismatch or "dysfunction" of climate knowledge with climate reality.

So, sadly, it appears that mainstream climate science models, infected by leftist politics and ideology, are currently experiencing an embarrassing condition properly diagnosed as Climate Dysfunction or CD. This condition can very well be linked to Low T (lower global average temperatures than predicted). (Note that CD is not to be confused with RD or Reptile Dysfunction, which is easily cured by provision of a new terrarium.)

The actual global average temperature trend has been rather flat for quite some time (about 18 years, in fact). The problem is that this flaccid condition was not at all predicted by vaunted climate models.

Now, before panic sets in and all kinds of pusillanimous explanations are given for the impotent performance, try revitalizing the climate change hypothesis using remedies from the youthful days of climatology—days when humility was more in vogue. Renewed vigor might include more emphasis on issues such as the complex role of water as a climate regulator and the sun's contributions (including solar variability and "cosmic-ray theory"), and generally allowing reasonable challenges that bring stimulating ideas to the ongoing research (rather than simply dismissing challengers with schoolyard name calling).

Finally, a warning: For CD lasting more than 14 years (which it already has), seek attention to reality... Immediately!

In the Land of the Midnight Sun

EVEN IF YOU have never beheld majestic Mt. McKinley (which may soon become a politically incorrect moniker) or worked along the expansive Alaskan pipeline, you can appreciate the natural beauty and natural resource benefits of the magnificent state of Alaska. As I reported in the Preface, during the summer of 1977, I worked in Alaska, 160 miles above the Arctic Circle, as a weather observer on the shores of the icy Chukchi Sea, where the sky is lower, the air colder and crisper, than almost anywhere else on earth.

Unfortunately, I didn't stay long. After a short stint as a weather observer, recording snowfall in July and awesome displays of the aurora borealis, I returned to the Midwest and western Pennsylvania to pursue a career as an air-pollution meteorologist.

Although not as grand as observing weather from the top of the world, the air-quality profession is certainly an exciting, important field, especially today, with the anxious balance between energy needs and a clean environment. And, regardless of the storylines ballyhooed at the innumerable U.N. climate-change conferences, the balance is possible, of course, thanks to phenomenal advances in natural-resource extraction and contaminant-control technologies.

Alas, though, too much international fretting is abridging our connection to a secure energy future—a future with exceptional promise.

Natural resource exploration and development are key to independence. But it's more than merely energy-independence. The U.S. could become a major supplier of fuel to the rest of the world.

So, I long to see truly wild and wonderful Alaska continue to prosper as not just a terrific employment or tourist destination, but as an even bigger energy lifeline to the lower 48 states and the world, as well.

But, there is considerable difficulty in tapping Alaska's frozen frontier for its copious energy reserves. The difficulty is not so much physical as it is political, and the politics are sometimes laundered through the U.S. courts.

In 2013, Royal Dutch Shell was gearing up to continue conducting responsible exploratory drilling in the Chukchi Sea off Alaska's northwest coast later in 2014. However, the company had to postpone its plans (now apparently a go for 2015) because of a decision by the Ninth Circuit Court of Appeals. In a January 30, 2014 conference call with investors, Shell's Chief Executive Ben van Beurden said that the company was:

> *...frustrated by the recent decision by the [court] in what is a 6-year-old lawsuit against the government.... The obstacles that were introduced by that decision simply make it impossible to justify the commitments of cost, equipment, and people that are needed to drill safely in Alaska this year.*[x]

Senator Mark Begich (D) said of the court decision:

> *It is simply unacceptable that judicial overreach is getting in the way of letting Alaskans develop our own natural*

---

[x] *Oil & Gas Journal*, January 30, 2014

resources.

More recently, a lawsuit was filed by the state in the U.S. District Court for the District of Alaska challenging the U.S. Fish & Wildlife Service and Department of Interior because of their rejection of Alaska's plan for oil and gas exploration in a portion of the Arctic National Wildlife Refuge.

On March 14, 2014, Governor Sean Parnell (R) announced that:

> *It is both disappointing and disturbing that the Obama administration, which claims that it is pursuing an 'all of the above' energy policy, is afraid to let the people of the United States learn more about ANWR's oil and gas resources. The modern technology that we are seeking to use [including advanced three-dimensional seismic imaging] is responsibly utilized all across the North Slope with extremely limited environmental impact, and would dramatically improve our understanding of ANWR's resources....*

Phenomenal progress in natural-resource exploration and extraction and contaminant-control technologies has afforded us a successful balance between energy needs and a clean environment. The U.S. has been the world's leader in such energy-resource activity, and our supply is plentiful.

The Department of Energy ranked us as the world's largest petroleum and natural gas producer. Furthermore, the International Energy Agency predicted in November 2012 that the U.S. would be a net oil exporter by 2030. In addition, with the expectations of a dominant percentage of the world's oil shale energy reserves added to our leadership in coal and advancements in biofuels, nuclear power, and solar and wind energy, we could readily become a major supplier of fuel to the rest of the world, including the EU in its time of strategic need.

The U.S. has the rich resources and the sophisticated and ethical ability to do right by the planet and its people. Does it have the political will? What happens in the near future in the Alaskan chilly clime of spectacular beauty and abundant assets could be the bellwether.[xi]

[xi] As this book was going into production, the front page of the September 29, 2015 issue of *The Wall Street Journal* announced that "Shell Drops Arctic Oil Exploration." The news item by Sarah Kent reported that "Royal Dutch Shell PLC is quitting its $7 billion Arctic campaign after drilling just one well with disappointing results, becoming the latest big oil company to abandon the riches under the northern seas in the face of stubbornly low crude prices."

Does Progressivism Mean Progress?

DURING MY THIRTY-FIVE years as a government scientist, industry consultant, and academician, I have witnessed the increasingly adverse influence of progressivism on science. This influence has been especially felt in the environmental sciences.

From the start of the modern environmental movement with the publication of *Silent Spring* by Rachel Carson in 1962 followed by *The Population Bomb* by Paul Ehrlich in 1968, the environmental sciences have become highly politicized and combative.[84] Differences of opinion are fine and even desirable in science. But progressives have brought dogma, orthodoxy, and conformity to the environmental sciences. Anyone who opposed the idea that the Earth needs to be protected from humans no matter how much it costs was treated as not merely an intellectual opponent, but as an enemy. That justified the use of ad hominem attacks, public ridicule, campaigns to deny funding, and even eco-terrorism.[85]

Such tactics would not be effective were it not for the fact that the media, academia, and the broader scientific community tolerate and in some cases even encourage them.

For example, the response to the Climategate scandal revealed tremendous bias, hypocrisy, and indifference to words and actions that would otherwise be considered highly improper—if not worse.

By any reasonable standard, the Climategate scandal should

have been a major setback, discrediting both the AGW theory and its proponents. Instead, the media and the science establishment worked overtime to minimize the damage caused when embarrassing e-mail messages and documents were made public. The media downplayed the importance of the content that was disclosed and referred to the files as "stolen," implying that the hacker committed a crime. Indeed, accessing and exposing private e-mail messages probably was a crime. However, the same news organizations usually treat anyone who accesses and discloses information damaging to large businesses or conservative politicians as a courageous whistleblower. The perpetrator is excused for stealing private information on the basis that he or she performed a service to society.

Similarly, the science establishment circled the wagons, vilifying anyone who suggested the need for a serious independent investigation of the researchers. Yes, there were some investigations. Unfortunately, these probes were conducted by universities, government agencies, and organizations that tend to benefit from what has become the AGW industry. This was analogous to appointing groups with ties to the oil industry to examine the causes of and responses to the Gulf oil spill. Naturally, such groups would be inclined to exonerate BP. When the investigations concluded that the e-mails and documents did not reveal misconduct on the part of researchers—despite talk of suppressing evidence and conspiring against dissenting scientists— neither the media nor the science establishment questioned the findings.

An ideology such as progressivism is not automatically bad for science. There's nothing wrong with researchers focusing on scientific issues that could help or harm society. The problem occurs when they let their ideology influence their scientific judgment. The risk of that happening is significant, because political ideologies typically claim to explain and solve a wide range of problems. When scientists try to make their experiments, research, and theories conform to a political ideology, the result is usually bad science. (The Lysenko affair, in which the Soviet Union banned the study of genetics, is one of the more infamous examples.)[86] The natural world is complex and often surprising, and good scientists understand that it's never wise to rule out possibilities for non-scientific reasons. If the history of science teaches us anything, it is to expect the unexpected. Scientists who are guided by ideology may try to make everything conform to their preconceived notions.[87]

Still, different perspectives are beneficial to science and can even be rejuvenating. Progressivism is destructive only when it tries to control the way that scientific concepts are proposed and evaluated.[88] Unfortunately, that's exactly what appears to have happened in the environmental sciences.

When progressive environmentalists warn of the "dangers posed by unprecedented, uncontrolled, and unchecked human alteration—be it biological, chemical, or physical—of our environment,"[89] the science establishment accepts it without questioning. However, when conservatives warn of the dangers posed by a big, unchecked, and intrusive federal government, not only is their knowledge of science ridiculed, but their motives and overall intelligence are questioned.[90]

What will environmental science look like thirty years from now? Let's hope that the dominant ideology still includes *hypothesis, experimentation,* and *observation* and is not simply conformance to progressivism.

SideBar32: Backward to the Future

The message is clear from the mounds of People's Climate March literature littering the landscape—"progressives" want to take us back to the "good old days," to the days when life was brutal, banal, and brief. Days when governments ruled with an iron fist, wars for territory and religious domination abounded, and humans were at the mercy of nature.

Perhaps the fastest way to achieve this sort of progress is to force retreat to the most ancient of energy sources—sunbeams and breezes. Ignore true advancements in energy extraction from heretofore unattainable sources like the atom and mile-deep fuel deposits.

If the numerous gigantic wind propellers impaling the landscape and the Ivanpah solar-electric generating plant brilliantly blotting the Mojave Desert are any bellwether, the progressive energy future belongs to the birds: sliced, diced, and deep fried.

It turns out that as wind turbine blades cut through the air, they also do the same to birds, including the endangered kind. Ditto on the gargantuan Mojave sun catcher, down go the feathered flyers in flames.

Despite the unintended consequences of breeze and beam boondoggles and propaganda pandered by progressives, the United States is still the most tolerant and technologically advanced nation on earth in history. The freedom to responsibly extract and develop modern energy sources will only increase our power, independence, and influence to be a force for good across the globe.

This force for good is desperately needed right now with the world so bent on returning to the "bad old days":

With governments (including our own unfortunately) vying for more control over the lives of its citizens;

Religion harking back to pre-Medieval days as impious, pitiless punks impose free-choice conversions—of the convert-or-die variety—on innocent victims;

And, of course, nature doing what it does naturally, making life as challenging as possible for everyone.

But, by progressive environmental religionists crusading for phantom issues like "climate justice," attention and finances are being diverted from the truly desperate situations in the world.

More attention and money should be paid to the critical conditions like Russian, Chinese, and Iranian aggression, and ISIS.

Furthermore, those with true concern for the environment should maintain a broad perspective. For example, investing in the fight against terrorists protects the environment in a big way and pays huge dividends. After all, if ISIS finds human life so expendable, it probably won't give a hoot about preserving the spotted owl.

So what's a better alternative to marching against the ethereal enemy of manmade climate change? Baylor University professor Rodney Stark provides a solution.

In his latest book *How the West Won: The Neglected Story of the Triumph of Modernity* (Intercollegiate Studies Institute, 2014), Dr. Stark aptly notes that Western modernity, even with all its limitations and discontents, is still "far better than the known alternatives—not only, or even primarily, because of its advanced technology but because of its fundamental commitment to freedom, reason, and human dignity."

Freedom, reason, and human dignity, now those are conditions worthy of a commitment to worldwide dissemination.

Why would we want to return to our ancestor's life that was severe, although, at least it was short? Like Woody Allen's classic joke told in *Annie Hall*, "the food at this place is really terrible... Yeah... and such small portions."

With greatly prolonged life expectancies today and with so much potential for expanded comfort and convenience for people across the planet, we could all experience good food and large portions. But, progressives are working tirelessly to march everyone backward to the future—a future once again likely to be brutal, banal, and brief.

Saul Alinsky, Climate Scientist

IN HIS 1971 handbook, *Rules for Radicals: A Pragmatic Primer for Realistic Radicals*, the godfather of community organizers, Saul D. Alinsky asserted that the "basic requirement for the politics of change is to reorganize the world as it is."

To Alinsky, the world and its history were all about revolution.

Saul Alinsky radicals who are all about revolutionary change ("we are the ones we've been waiting for" kind of change) have now seized control of an issue that can more quickly bring about that change—"climate *disruption*," as expressed in community-organizer lingo.

If Alinsky were alive today, he would likely fit right in with the current activist climate scientists. Alinsky would probably see that the challenge is to convince enough of the "Have-Nots" that their privation stems not just from racism, sexism, classism, and all the other social -isms that are used to divide people, but also from what some have called "climatism."

Ensconced in political power, today's Alinsky-style radical elites running roughshod over pure scientific practice can force societal change predicated on unfounded predictions of climate doom. They seem willing to use any means necessary to realize their society-remaking goal. After all, to such radicals, the ends justify the means.

And, it looks like that in their struggle to amass more control over the masses, climate activists are using tactics straight out of the Alinsky playbook.

Take the first rule for radicals: *"Power is not only what you have but what the enemy thinks you have."* The simple fact is that the evidence touted in favor of harmful man-made climate change is inconclusive at best. But you wouldn't know this if you follow climate activists' unsubstantiated, arrogant assertions. The activists make believe their conclusions are incontrovertible; they want people to simply "trust them, they're scientists."

Whereas the second rule cautions *"Never go outside the experience of your people"* in order to avoid your own cohort's "confusion, fear, and retreat," the *third* rule advises: *"Wherever possible go outside of the experience of the enemy,"* to cause "confusion, fear, and retreat" in your opponents. Climate activists, along with their shills in the media, are doing just this to prominent politicians who dare to question human's substantial contribution to climate change. And, unfortunately, some of these politicians, because of their own ignorance, are doing nothing to help their cause. Which eventually leads to...

"Ridicule is man's most potent weapon," apparently a favorite tactic of climate activists. It's far easier to submit a barrage of ridicule or to sling labels like "denier" at people, than to engage in thoughtful scientific debate—especially if the facts are not in your favor.

Skipping along to the seventh rule, Alinsky notes that *"A tactic that drags on too long becomes a drag"* ...or, in the case of "global warming," it gets proven wrong. Climate activists started with hyping dire predictions about global warming and had plenty of computer models to back them up. When real life showed those models to be spectacularly wrong, the activists embraced the term "climate change," to continue hoodwinking the unwary public.

Real trouble for the truth is found in execution of the ninth rule: *"The threat is usually more terrifying than the thing itself."* For example, there had been a lot of reporting in early 2015 on

Arizona Rep. Raul Grijalva—in his own words, "a Saul Alinsky guy"—and his recent attempt to collect damaging information on several professors and climatologists who don't toe the leftist line on climate change. How many other scientists, interested only in actual non-political climate research, are now going to be more acquiescent to the climate activist position out of fear of similar attacks? Or, how many advancements will go undiscovered because good scientists don't want to get caught accepting funding from politically-incorrect sources?

In the thirteenth and final rule—probably Alinsky's most well-known and well-worn rule—radicals are directed to *"Pick the target, freeze it, personalize it, and polarize it."* Climate activists, especially when confronted with inconvenient facts, like to target the messenger rather than prove their own guarded hypotheses. Assaults are carried out with *ad hominem* attacks on legitimate challengers (aka, "deniers"). Then they are linked to some perceived hobgoblin like "big oil" or "the Koch brothers."

Alinsky would have been proud of climate scientists operating essentially as social activists. Not only do the activists get to liberally spout their ideas, they also get to force them on others. Activism can be used to mask intolerance, in this case intolerance to dissenting voices in atmospheric science.

So, the real challenge is for reasonable dissenting scientists and engineers to convince ordinary, good, decent Americans that, since the publication of *Rules for Radicals* in 1971, their misery stems largely *not* from racism, sexism, classism, or even climatism, but from the execution of the most pernicious -ism of all—socialism.

SideBar33: More Insight Inspired by the Alinsky Manifesto

Any revolutionary change must be preceded by a passive, affirmative, non-challenging attitude toward change among the mass of our people. They must feel so frustrated, so defeated, so lost, so futureless in the prevailing system that they are willing to let go of the past and chance the future.

The problem for the current Administration is that they have inadvertently generated the former without producing the latter.

Furthermore, after six years of the Obama administration forcing all its leftist changes on the U.S. public, now, by Alinsky's definition, it seems that today's ruling-class radicals are the new "Haves." "The Haves want to keep things as they are and are opposed to change. Thermopolitically they are cold and determined to freeze the status quo." Thus, our betters in politics, academia, and the media apparently want to keep U.S. citizens separated by class and by race, to keep America subdued on the world stage and ineffective for maintaining good and order across the globe.

Overall, the Administration seems to be acting on the leftist tact that dictates that "the ends justify the means" when it claims that the limits of carbon dioxide can be brought about largely through energy efficiency measures. However, as a good general principle, taking the right action for the wrong reason is wrong, since bad things can happen. These bad things have manifested themselves as over-bearing government control over the economy and bad judgment on natural resource projects (for example, the federal government's multi-year practical squashing of the Keystone XL pipeline, a project that looks to be a conduit for continued energy security, efficiency, and safety).Furthermore, the ends-justifies-the-means tactic often leads to unintended consequences. To avoid such consequences, the government should try doing the right thing in the right way for the right reasons at the right time.

Alas, the Alinsky/leftist ideology ultimately simply deconstructs to a system promoting profound arrogance.[xii]

[xii] Although a lack of humility is likely a big issue for Alinsky/leftist ideologues, such narrow-minded thinkers might counter with the old saw, "It's hard to be humble when you are as great as I am."

Enlightened Activist Scientists Dim Society

PROGRESSIVE ACTIVISM HAS been taught by our schools for so long that it has become part of the US's culture. And it's making its mark and taking its toll.

To some people, an "activist" is someone with social concerns who has decided to do something constructive. To others, an activist is someone who tries to impose their ideas on everyone else. Activist judges reach beyond existing laws to achieve their desired goals. Activist legislators propose and vote for laws that force people to make what the legislators consider the right choices. Activist officials impose their vision of a better society through ambitious programs and extensive regulations.

Is an activist scientist just a scientist with social concerns who has decided to do something constructive? Or is an activist scientist a person who believes they possess greater knowledge and, therefore, are justified in imposing their ideas on those who are less knowledgeable, such as the average non-scientist?

How did the activist scientist emerge? The 20th century was a century of great scientific discoveries and technological innovations. Scientists came to be revered in Western societies. In fact, science has accomplished so many good things that some imagine scientists can solve all of society's problems. This belief has been reinforced in many ways: the accuracy and rigor associated with science, the rise of the information society, and the popularity of scientific methods in the social sciences are just a few examples.

That may explain why so many scientists have become activists, but why do they consistently oppose business, industry, and the military? In part, it may be because scientists believe they deserve greater fame, fortune, and power. And it may be that the success of Western countries has left many scientists (as well as others) feeling guilty. And there's the simple reason that those who warn of large-scale dangers attract funding and publicity. But there's also an element of intellectual laziness: It's easy to criticize those who work in other fields, but not as easy to offer alternatives that are both superior and realistic in their fields.

For whatever reason, the idea has taken hold that people are destroying the planet and time is running out to rescue the Earth. And this idea, which depends more on opinion than fact, has become a major focus of education. Certain fields have become popular in college and graduate school: social work, education, health care, and environmentalism. Young people are taught that helping others is a noble endeavor, which is certainly true, and many pick what appear to be the most altruistic careers. Any student focusing in one of these areas will get plenty of encouragement as well as the resources needed to fight the good fight. The one thing that these students typically do not get, however, is a different perspective on environmental issues. Engineering (including environmental engineering) is an obvious exception, because engineers must learn how to approach every problem or challenge from a practical perspective.

In general, environmental studies tend to be based on qualitative analysis and, therefore, environmental policies are more vulnerable to hand-waving arguments, hyperbole, and unsupported extrapolations. Plus, fields that don't depend as much on hard numbers are more susceptible to emotional arguments. Graduates who major in environmental studies tend to view the world and humanity's place in it as they are taught. Consequently, they think in terms of the dangers emphasized by their professors and textbooks, and find it difficult to think in terms of the thoughtful use of resources and actions that could benefit both humans and their surroundings. Most graduates are, at least in mindset, minions of the activist progressive environmental groups that are fueled largely by political wood, hay, and stubble. It's hard to think in terms of protecting and nurturing humans when you've been told over and over that humans are damaging and possible destroying the planet. Meanwhile, we are assured by an elite group of mainly academic scientists that disaster awaits us if we don't take drastic and enormously expensive measures to change the future course of the Earth's climate. However, the global climate 100, 50, or just 10 years out is unknowable at any practical level. The climate forecast maps—including those presented in IPCC reports—that depict future temperature and precipitation conditions are pure speculation. Yet billions of dollars are spent not only to research future conditions (which is a reasonable endeavor to the extent that it increases our knowledge) but to reshape society. And it's fueled by a gigantic financial feedback loop: funds are dispensed to researchers who make dire forecasts that justify the need for more funding.

Activism in the name of science takes a heavy toll on society. It corrupts science, because it encourages scientists to seek evidence and develop theories that support their political causes, and it discourages other scientists from presenting evidence and developing theories that are likely to provoke hostility. And it corrupts society—both by depriving it of scientific freedom and by creating huge science boondoggles. Trillions of dollars may soon be given to the United Nations and associated groups because there are people who deny the mysteries of nature and tout their own brilliance. However, if these science luminaries are brilliant, it's only because they have successfully conspired with politicians and bureaucrats to create an entire industry based on climate fortune-telling.

SideBar34: The Politics and Wishful Thinking of Climate Predictions

Long-term climate predictions may be more about politics and wishful thinking than science. And, climate-change confusion by the general public and the potential benefits this confusion can reap for ruling class opportunists continues to loom large on the horizon.

Right now, a big concern in the atmospheric science community is the fact of a global warming "hiatus." For more than 18 years, average global temperatures have remained rather steady, despite confident climate predictions to the contrary. Sophisticated climate models developed and run by the most expert researchers in the field spewed forth reams of details on a future warmer world. Yet, in reality, what transpired was a statistical flat-line, nada, a no-show on a Gaia-the-hottie.

Not to worry, though, we're assured that the planet will surely heat up, unless middle-class Americans in particular acquiesce to their betters in government and academia and stop using so much fossil-fuel derived energy.

But, in more reasonable times, with so many urgent issues to address like deadly disease, terrorism, and tyrannical imperialism, the powers-that-be would welcome a reprieve from what was forecasted to be an imminent disaster. Perhaps, even if a disaster eventually results from our affordable, comfortable living, at least for now we have time to deal with the more pressing worries.

But why doesn't reason prevail? Why do calls for cuts in "carbon pollution" become more shrill, and warnings of climate doom more dire?

For the answer, look at the situation from the pompous

politicians perspective. Here's an issue with lots of room to stretch the truth while conducive to tremendous emotional appeal like save-the-future-for-our-children kind of appeal. For a potential tragedy pending so far in the future, the consequences of "doing nothing" are limited only by a wild imagination. The taxpayer dollars directed to a dubious atmospheric appraisal is a lawmaker's dream for raking in the dough to maintain control, direct money to pet projects, buy votes, or just keep campaign contributors flush with cash.

On the other hand, wishful thinking is the purview of academic theoreticians. On the cloistered college campuses across the country, theory trumps reality. So, if real-world global average temperatures have departed from what *ex cathedra* scholars divined through climate models, guess what needs to be aligned.

Climate models, which have been largely conceived and nurtured at the universities, are impressive tools of the trade in atmospheric science. They are remarkable achievements in our understanding and practical simulation of a complex dynamic system. But, when the predictions of the models don't match reality, it's not the reality that needs adjusting.

Yet, many bullying tactics have prevailed from the schoolyard to maintain the authority of the academy. Perhaps the most persistent is name-calling. But, if a research scientist uses the term "climate change denier" or, even sillier, "climate denier," to defend their territory, they are not just revealing their immaturity, they are also being dishonest. And, dishonesty has no place in good scientific practice which requires integrity.

What about the claim that the debate is over and the conclusions are settled? As a veteran atmospheric scientist, I know there never was a rigorous "debate" over whether humans are substantially responsible for long-term, global climate change, so the debate can't be over. Furthermore, the science is still very much in a developing stage, so it certainly can't be "settled."

What is needed now is a hiatus in collegiate rhetoric, and a hiatus in government effort toward the non-issue (or at best the small problem) of man-made global warming. More focused attention must be directed toward large problems, such as the ever-expanding terrorism. If not, so many more innocent people may never live to see any kind of future climate, hot or not.

Climate Disruption Drivel

WHEN OCCUPY WALL Street took center stage by restricting the public's access to streets and parks, global warming was temporarily overshadowed. But that did not mean an end to meteorological mischief. The political rhetoric blaming climate change on the industrialized nations of the world has since become even more biting and aggressive.

Rather than strengthening the foundations of climate science by releasing raw data for third party evaluation, presenting their hypotheses for open debate, and interpreting the data in a transparent manner, government-funded scientists instead decided to enhance the way they frame the issues and deliver their message. The latest tactic embraced by manmade global warming proponents is to sharpen their message and increase their influence over the editorial boards of environmental journals. The goal is nothing less than to deflect, discredit, and denounce challenges to the climate catastrophe canon.

There's nothing wrong with scientists getting a little help in presenting their ideas clearly and effectively. But it's worrisome when scientists enlist public relations experts and linguists to help them plan and carry out propaganda campaigns. For example, in a special workshop at the December 2010 American Geophysical Union (AGU) meeting in San Francisco speaker after speaker advised attendees about how to prosecute a successful war of words. The theme of the January 2011 annual meeting of the

American Meteorological Society was "Communicating Weather and Climate," echoing the tactics introduced at the AGU gathering. The following month, the National Center for Atmospheric Research hosted a seminar in Boulder, Colorado to help climate scientists with science communication issues. The speaker was the author of The Republican War on Science—a book brimming with politics but almost completely devoid of science.

Anyone can abuse science—whether progressive, conservative, Marxist, or libertarian. And it should be clear that politicizing science is one of the most egregious abuses. The purpose of science is to uncover the truth. To do that, scientists must leave their assumptions and biases behind. They should base their findings on experiments that are verifiable and repeatable. If one scientist discovers that mixing two chemicals together produces a violent reaction, then other scientists should be able to perform the exact same experiment and get the exact same results. It shouldn't matter whether they are Republicans or Democrats.

The difficulty arises when scientists try to explain the facts by developing theories. There's nothing wrong with proposing theories—good theories lead to further discoveries—but scientists must be careful not to confuse theory with fact. Theories can and should be contested. This is why science is respected. It draws a bright red line between facts, which can't be argued, and theories, which can and should be.

Some theories may appeal more to progressives and other theories may appeal more to conservatives. But good scientists will be careful not to let their political views influence their scientific judgment. Launching a strategic PR campaign to promote a theory that is also a political cause is unlikely to advance unbiased scientific understanding or practices.[91]

Integrity in science requires clearly stated hypotheses, reproducible results and data, and mathematically rigorous interpretations. Science doesn't condone bending the data to favor personal beliefs or causes. Nor does the unvarnished truth require political spin. Science shouldn't be put up for sale, either. Spinning tentative and weak science to make it look definitive and strong is not part of the scientific method.

Nature is the ultimate teacher. For scientists, humility and a willingness to go wherever the data leads are fundamental virtues. Arrogance is especially dangerous because it can blind researchers to the possibility that their cherished hypotheses are inconsistent with reality.

Science is never settled. It is a never-ending journey of investigation. Hypotheses are proposed and then data are gathered and analyzed in an effort to prove or disprove them. Climate investigations are particularly endless, because the climate system is inherently complex, global in scale, and affected by both internal and external factors. Assertions by any individual or group about the nature and causes of long-term, global climate change should be viewed as what they are: speculative.

Convincing all or most scientists to acknowledge and respect the line between facts and speculation isn't going to be easy. In James Lawrence Powell's book, The Inquisition of Climate Science, AGW proponents are portrayed as the ones who are under attack. Ironically, any knowledgeable scientist who challenges the current climate science dogma could write this book almost chapter for chapter. Even more ironically, despite the fact that AGW proponents get the lion's share of funding, support from the education establishment, and favorable stories in the media, The Inquisition of Climate Science complains that not enough of us accept the claim that our lifestyles are responsible for dangerous climate change. The author asks, "Why, when the scientific evidence for global warming is unequivocal, does only half the public accept that evidence?" Dr. Powell goes on to conclude that "in the denial of global warming, we are witnessing the most vicious, and so far most successful, attack on science in history." (By "global warming," Dr. Powell means human caused global warming.)[92]

It is amazing that even someone with a long list of scientific credentials misses the point that science should welcome reasonable challenges to the *status quo*. Instead, the author is exasperated by continuing challenges to the dominant perspective. In that sense, *The Inquisition of Climate Science* is typical of books written by those deeply immersed in the science establishment. They believe that the establishment view is obviously the correct view, and that dissenting views are at best a nuisance.

Like other books defending AGW, *The Inquisition of Climate Science* depicts opponents as shills for Big Oil and Big Tobacco.[93] There's no admission that government and the science establishment could likewise be labeled "Big Government" and "Big Science." As the saying goes, if you want to identify the real movers and shakers, "follow the money." In 2011, the federal budget proposed allocating $2.6 billion for the Global Change Research Program.[94] More recently, the Obama administration asked for $770 million to fight global warming in developing countries.[95] No wonder anthropogenic global warming attracts such ardent support: US government funding, alone, is sufficient to employ at least 30,000 researchers and administrators.

Dr. Powell cites the testimony of groups including the Center for Media and Democracy and Union of Concerned Scientists as if everyone agrees that these groups are balanced, unbiased, and universally respected. The Center for Media and Democracy describes itself as "exposing corporate spin and government propaganda," but it clearly focuses on opposing Republicans. The Union of Concerned Scientists, which some may mistakenly believe is interested in fairness, is very concerned about environmental issues, but apparently only for the purpose of advancing their progressive agenda. Further dispelling the notion that AGW proponents are the ones under siege, Dr. Powell cites Greenpeace, a multinational non-governmental organization with operations in more than 40 countries. Complaining that politicians are the group most often quoted about global warming, based on a media study by Maxwell T. Boykoff and Jules M. Boykoff published in 2004,[96] Dr. Powell missed the irony that one of the loudest and most effective voices promoting climate change hysteria since the 1990s is former Vice President Al Gore.

Despite all of this, Dr. Powell laments that "our political leaders are unwilling to get ahead of public opinion." In other words, he believes that politicians should impose certain beliefs on their constituencies rather than respect the will of the people. Continuing, he hopes "public opinion can change if people will think for themselves and look at the evidence for global warming and at the deceitful and mendacious claims of the industry of denial." So there's the rub: People are not thinking for themselves. Unfortunately, progressives often make their case by accusing opponents of being either ignorant, malicious, or both. And they often get away with it. For example, when the public doesn't share a progressive view, then it must be because shadowy forces have obstructed those trying to enlighten the masses. It never occurs to them that people who think for themselves might reject progressive views. After all, what good is thinking for yourself if it does not lead to deciding for yourself?

SideBar35: 5 Moves in a Rigged Climate Crisis Game

As the President continues to deal out more rounds of anthropogenic climate-change hype against an ethereal enemy— i.e., reliable, inexpensive energy—claims of humans caused this or that meteorological mayhem is sure to follow.

Climate activists apparently have discovered an effective scheme to constantly draw an association between noteworthy weather and human culpability. That dubious association comes in just five easy moves:

Find a serious, recent weather event.
Figure out what was unique about it.
Claim that uniqueness was because of human activity.
Get the media, politicians, spin doctors, pastors and pontiffs, Hollywood celebrities, and socialites and socialists of all stripes to vent your discovery of man-made disaster.
Then wait for fame and fortune to fall from the frenzy.

A manufactured climate crisis is easy when you stack the deck in your favor.

Countering AGW Propaganda

WHILE THE GLOBAL warming elites enjoyed the good life at the UN's 17th Conference of the Parties (COP) in the resort-like city of Durban, South Africa, the rest of us were busy wrestling with more urgent and pressing problems. The American economy is struggling with weak markets, high unemployment, and tremendous financial uncertainty in a political climate that is increasingly hostile toward businesses.

Despite the fact that individuals and society benefit greatly from industry, industry leaders are almost always painted as the bad guys by environmentalists, and they get only slightly better treatment from the media. Whether it's the socialites splashing in the South African sun, the students and professional protestors congregating on streets and in parks under the Occupy Wall Street banner, or others decrying income inequality (and whether they admit it or not, advocating socialism), the common message seems to be that the greatest producers among us—the industrialists—are somehow out to enslave the workforce in order to enrich themselves. In a sense, manmade global warming is simply the latest version of class warfare.

With the fall of the Soviet Union and China's ostensible embrace of free market principles, traditional socialism and communism have been discredited. However, manmade global warming and opposition to Wall Street greed provide new vehicles for those who refuse to give up on the dream of a top-down, classless society run by benevolent autocrats.

Industry has had a rough time trying to deflect and counter accusations coming from multiple quarters. Efforts by major corporations to demonstrate that they, too, are interested in conservation and a safe environment are generally dismissed as half-hearted or mere talk.

However, industry can help its cause by embracing several basic principles and practices. I and my co-author, Mark D. Shull, have detailed these in *Environmental Risk Communication: Principles and Practices for Industry*.[97] The key elements of an effective program to counter organized opposition to industry include:

1. Operate legally and ethically—Facility owners and operators must do more than just comply with the myriad laws and regulations. They must consistently do what is right. Treat employees and the community in a way that is both fair and just.

2. Educate employees about the benefits and the risks— Employees should be well informed about the benefits their company provides to the community. In addition, they should be familiar with the risks and environmental impact associated with the company's operations as well as ways to mitigate those risks. Remember that every employee is both a productive member of society and a potential ambassador to the local community.

3. Listen and respond appropriately to the public—Effective communication requires trust, and trust is built on an honest, open dialog between all parties affected by plant operations.

4. Promptly disseminate information to the media—Proactively disseminate clear, accurate, complete, timely, and consistent information about your facility to head off disinformation campaigns by activists and biased reporting by the media. Unfortunately, news outlets often resort to story-telling when

there is a shortage of good information, so it's important to give them the facts and access they need to do their jobs properly.

Diligently practicing these four guidelines should help convince levelheaded people that the operation is beneficial and reasonably safe. However, with today's Internet, negative perceptions can spread at the speed of light, so maintaining positive community relations has become a 24/7 job. Beware that anti-industry zealots will not be convinced and never tire of attacking industry.

Some industries need to be more vigilant than others. These include nuclear power plants, chemical processing plants, and hazardous waste treatment facilities. Lately companies engaged in extracting, processing, and using fossil fuels have been the primary targets of activists, particularly since carbon dioxide has been legally declared as a hazard to the planet. Though coal and oil have been vigorously attacked—witness the ongoing campaigns against Canadian tar sands and the Keystone XL pipeline[98]—the heaviest fire is currently being directed against a relatively new villain, natural gas extracted using fracking technology.[99]

There are certainly potentially safe techniques for extracting natural gas from the Marcellus and Utica Shale formations, and these could go a long way toward providing long-term energy security for the US. In fact, the US is estimated to possess nearly 80 percent of the world's shale-based energy reserves. By adding such a huge amount of natural gas to an already plentiful supply of coal and oil, the US could become not merely energy independent, but a major exporter of fuel to the rest of the world.

Even that seems to disturb environmental activists. Many simply don't want to see the US achieve energy independence. Instead, they want the US to become just another member of the international community, operating under the supervision of the United Nations. Ironically, the demise of the Soviet Union has had one ill effect: it has removed from the scene a vivid example of the incompetence, corruption, and brutality of massively centralized power. Progressive environmentalists, in opposing private industry, are pushing for an even greater concentration of wealth and power in the hands of people and organizations that are largely unaccountable.[100]

Pollution Prevention Practices for Industry

Industry can save money and reduce the amount of industrial waste it adds to the environment by following pollution prevention (P2) practices. P2 techniques have been around since the start of the modern industrial era, but only in the past few decades have the techniques been more formally recognized, categorized, and endorsed by a wide array of stakeholders including industries of all types, government agencies, environmentalists, and community residents.

A successful P2 program takes considerable research and planning, but it is well worth the effort. It can decrease operating costs, reduce the number of regulations that must be observed, lower the risk of accidents such as chemical spills, and improve relations with the local community.

The typical starting point for a P2 program is the waste audit. This can be performed in-house, though depending on the nature of the operation, it may require the assistance of outside specialists.

P2 practices generally include:

1. Initiatives to assure that top-level managers are committed to supporting the development of a plant-wide program with clear, achievable, and measurable goals.
2. Employee training to encourage their cooperation and input to the program.

3. Energy efficiency audits to ensure the best use of available power.
4. Preventive maintenance to increase equipment operating efficiency and to help avoid down time.
5. Improved housekeeping to keep the plant clean and free of obstacles that contribute to accidents involving employees and to prevent the release of materials to the environment.
6. Material substitution/reformulation to minimize the use and concentration of hazardous/toxic chemicals.
7. The three Rs of P2: Recycling, Reusing, and Reducing raw materials, process components, final products, and packaging.

Implementation of P2 practices need not be cumbersome; however, each item requires careful thought to reap the full benefit. For example, in addition to keeping shop areas tidy, improved housekeeping could include practices as simple as keeping lids on containers of volatile materials and products tightly closed to prevent fumes from escaping into the surrounding air.

SideBar36: Good Well Hunting

As part of my job as an air-pollution meteorologist, I have had the opportunity to go on-site at Marcellus shale gas wells in various stages of development and operation in southwestern Pennsylvania. The state-of-the-science technology and professionalism of the managers and operators were truly impressive.

While scoping out actual gas drilling in progress, it struck me why so many environmental activists are against such extraction activity.

First, activists apparently have no appreciation for and little understanding of the cutting-edge engineering involved with fracking—the technique employed to extract natural gas from areas roughly a mile below the surface. I and other environmental professionals with me were thrilled by what we observed of the drilling. Our backgrounds in science and technology, along with our many years of work in the real-world, provided us with an awesome perspective, much like what someone trained in the arts might experience at a rousing musical. (The analogy might appear to be a bit of a stretch, but I for one would rather watch a well-drilling operation than sit through a well-performed opera.)

Second, many are frightened by the unfamiliar—especially the unfamiliar that is also complex. It's been said, "Familiarity lessens fear." But, rather than become familiar with a modern, essential energy practice, many environmentalists would prefer to use their ignorant fear as motivation for their own actions and to enlist others to join their ill-informed timidity.

Third, after decades working closely with both energy professionals and environmental activists, I have observed that for the most part the former conduct themselves like anchored adults, while the latter act like jilted juveniles. From my personal observations, Marcellus well developers and operators are serious about running a safe, profitable business with as little disruption to property owners and the environment as possible, while their progressivist opposition will do whatever it takes to disrupt progress.

Regardless of anthems or antics, my job is to help quantify emissions and protect air quality from new Marcellus well activity. I can do both with a solid knowledge and appreciation of sophisticated technological operations such as fracking, and a reasonable concern for appropriate, careful use of the nation's ample supply of natural resources.

State of Fearful Climate Science

THE CURRENT STATE of climate science is worrisome. In his 2004 bestselling fiction, *State of Fear*, the late Michael Crichton introduced a skeptical climate science character, Professor Hoffman, who said "I study the ecology of thought.... and how it has led to a State of Fear." The professor went on to explain that the government practices "social control [which is] best managed through fear."

Could environmental activists, without intending it, be helping to prop up leaders whose rule thrives on fear?

In the nonfiction world, humanity is kept in a perpetual state of fear based on the latest IPCC talking points. Namely, people are being told that human-induced climate change is manifested in "extreme weather events."

In the U.S., the EPA is exploiting this alarming message in a uniquely creative way. The EPA is preparing guidelines for reducing indoor air pollutants in the expectation that people will spend more time indoors due to the increased amount of severe weather caused by climate change.[101]

This atmospheric angst is spread using deceptive tactics. First, an authoritative world body such as the IPCC reports an increase in severe weather and attributes it to anthropogenic global warming. Next, solutions are proposed that require altering lifestyles and shuttering coal-fired power plants. Services are offered such as education, research, consulting, and trading carbon credits. Oversight and enforcement are demanded of national and international bodies. Everyone seems to be cashing in on the doomsday predictions—private companies, academic institutions, and governments. Everyone wins—except for the losers who get stuck with the bill (the middle class) and the world's poor, who always seem to miss out on these massive wealth transfer schemes. Science also loses.

The creation of a single international body chartered to issue final decisions about such a complex issues is both arbitrary and arrogant. And it is certainly not consistent with the best scientific practices, under which anything short of a plainly demonstrable fact is considered tentative. This arrangement also invites corruption because it enables a few dominant personalities to impose their views on other members.

A much-needed exposé of the IPCC is provided by investigative journalist Donna Laframboise in her book *The Delinquent Teenager Who Was Mistaken for the World's Top Climate Expert.*[102] Ms. Laframboise observes that it is:

> both peculiar and ironic that an organization that so vigorously claims to represent a worldwide scientific consensus has systematically 'disappeared' so many consensus views held by so many different kinds of researchers.

She discovered that the IPCC "ignores the consensus among hurricane experts that there is no discernible link to global warming. The IPCC ignores the consensus among those who study natural disasters that there is no relationship between human greenhouse gas emissions and the rising cost of these disasters. The IPCC ignores the consensus among *bona fide* malaria experts that global warming has not caused malaria to spread." Ms. Laframboise concludes that in each case "the IPCC substitutes its own version of reality." A version that "makes global warming appear more frightening than genuine experts believe the available evidence indicates." [103]

To spread this committee-developed fear-mongering, people who know little or nothing about atmospheric science (such as politicians, entertainers, and spin doctors) are trotted out to testify in favor of the manmade climate change hypothesis, while everyday practitioners in climatology and meteorology who are skeptical of the IPCC's alarmist message are ridiculed.

Many experienced atmospheric scientists believe that scientists should be free to creatively apply their skills and perspectives to understanding the Earth's environment. Freedom is essential to discovery and innovation. Based on their actions, activist scientists believe that science is extremely fragile and must be shielded from dissenting views. Anyone who has studied the history of science knows that many great discoveries and theories were greeted with skepticism if not outright hostility because they contradicted the current consensus.

By maintaining a state of fear for the past few decades, climate science has strayed far from the scientific tradition of criticism and debate, damaging an honorable scientific profession. Worse, the state of fear discourages climate scientists from looking at things they are not supposed to look at, discussing things they are not supposed to discuss, and thinking ideas that are not supposed to be thought. That's not science—it's totalitarianism.

SideBar37: The Journalist and The IPCC, a bit about Donna Laframboise's 2013 Book, *Into the Dustbin: Rajendra Pachauri, the Climate Report & the Nobel Peace Prize*

About the worst thing that can happen to science is for politics to control it. And that seems to be happening in the field of climate science with the political culprit being the U.N.'s

Intergovernmental Panel on Climate Change (IPCC).

One enterprising journalist investigating this sad affair is Donna Laframboise. After her definitive 2011 expose on the IPCC, aptly called *The Delinquent Teenager Who Was Mistaken for the World's Top Climate Expert*, Ms. Laframboise has authored another gutsy critique, *Into the Dustbin: Rajendra Pachauri, the Climate Report & the Nobel Peace Prize*. Her new book— largely a collection of essays from Ms. Laframboise's blog NoFrakkingConsensus.com— is folksy in its style and compelling in its content. *Into the Dustbin* draws not only from statements in historic editions of the IPCC "climate bible" itself, but from direct quotes from the IPCC chairman, Rajendra Pachauri, and other "Nobel laureate" personalities.

Regarding the noble identification of such personalities, Ms. Laframboise notes that, because of the excessive and improper use of the Nobel laureate designation, a year ago the IPCC issued a statement about the 2007 Nobel Peace Prize that clarified that the honor "was awarded to the IPCC as an organization, and not to any individual associated with the IPCC. Thus it is incorrect to refer to any IPCC official, or scientist who worked on the IPCC reports, as a Nobel laureate or Nobel Prize winner."

Yet, as recently as this past April, Ms. Laframboise observes that the accolade continued to be bestowed even on minor individual authors of the IPCC climate bible. PR firms like Hoggan & Associates, who apparently specialize in disseminating anti- science and technology rhetoric, seem to be keeping the ruse alive by applying the term in press releases to one of their "Nobel Laureate" clients. (For more on products of Hoggan & Associates, see the DeSmogBlog.com website. Also see my review of James Hoggan's book *Climate Cover-Up: The Crusade to Deny Global Warming* in *The Washington Times* on June 4,

2010.[xiii]

Throughout *Into the Dustbin*, Ms. Laframboise documents the self-aggrandizing, political nature of the IPCC led by its zealous chairman, Rajendra Pachauri. Claims such as "the IPCC studies only peer-review science," "20-30% of plant and animal species [are] at risk of extinction," and global warming will continue to cause "more outbreaks of intense hurricane activity" confidently asserted by Pachauri and other IPCC-science promoters are deftly debunked by Ms. Laframboise.

Ms. Lafamboise disputes these claims noting that a "citizen's audit" of the 2007 climate report's 18,531 references discovered that "a full 30 percent (5,587) weren't peer-reviewed." The "20-30% extinction" value is preposterous as indicated in genuine peer-reviewed literature. And, the assertion that increased global temperatures will enhance hurricane intensity is outside of contemporary scientific understanding. Besides, recent history has shown the fallacy of increased hurricane activity with some of the lowest amounts and intensities of tropical storms on record.

Sadly, the final word, and often the only word, on climate change begins and ends with the IPCC report. The President, Democratic senators and representatives, U.S. Environmental Protection Agency heads, and so many others in government simply quote the IPCC to justify their power trips. This is why the revelations provided in *Into the Dustbin* are so urgently needed to be widely read. The book demonstrates the kind of clear, critical, independent thinking so essential for journalists and concerned citizens today.

The world is subjected to so much misinformation from politicians vying for control of the *hoi polloi*. And, the politically-driven global-warming "crisis" is just one of the convenient highways statists are speeding down to reach utopia for themselves

---

[xiii]   http://www.washingtontimes.com/news/2010/jun/4/book-review-climate-cover-up/?page=all

and their entourage. Hopefully, Ms. Laframboise's admirable efforts with *Into the Dustbin* and *The Delinquent Teenager* will be more road blocks than just speed bumps to our arrogant betters in the driver's seat of the climate-catastrophe bandwagon.

Perhaps her literary efforts can even help steer the leftist elites down the off-ramp and into the dust bin of history.

Meteorologists of the World Unite!

DESPITE CONGRESSIONAL REPUBLICANS' attempts to reign in the EPA, and a stay on more stringent ozone limits in 2011,[104] greenhouse gas restrictions for large power plants and other industrial facilities are currently being cast.[105] Consequently, US industries will continue to be shackled by bureaucratic chains forged with links connecting humans to harmful climate change.

However, given that no one can predict long-term, global climate change with certainty, is it reasonable to impose crushing regulations on today's industry? It is reasonable to continue studying whether the Earth's climate is changing over the long-term and whether human activity is a contributing factor. But taking drastic action today based on a consensus about what will happen in the distant future defies common sense.

That no one knows with accuracy and certainty what the atmosphere will look like in the future comes as no surprise to meteorologists. Any meteorologist who devotes a significant amount of time to forecasting quickly learns to be humble. The farther out the forecast, the more unreliable it becomes.

That applies not only to climate forecasts, but forecasts of all types. How many developments or trends can be accurately and reliably forecast decades in advance? The answer is few if any. The farther out the forecast, the more likely there will be major surprises.

My general rule-of-thumb is that for most areas of the US you can count on forecasts being accurate for up to approximately 72 hours. According to chaos theory, useful weather forecasts may look out about 14 to 28 days. So there is a period from about three days to up to four weeks at most during which weather forecasts are somewhat reliable. Beyond a few weeks, you might as well flip a coin.

The typical retort is that climate and weather are very different things. But that's a red herring. I would argue that predicting worldwide climate conditions decades into the future is an order of magnitude more difficult and unreliable than forecasting Wednesday's weather for Washington, DC on Monday morning.

The counter argument is that climate projections do not need to be as specific as mundane weather forecasts. True, pinpoint accuracy is not necessary for climate forecasting. However, even in sweeping time and space terms, the chances of successfully predicting the average temperature within a narrow range in the tropics or at the poles several decades from now are quite poor. Too many climate-determining variables are insufficiently known or unaccounted for, such as the impact of water in its three forms (vapor, droplets, and crystals) and solar activity.

As we saw earlier, the official US prediction for the hurricane season of 2006 was a bust. This wasn't a specific weather forecast—it was a general outlook for climate conditions over a region. It seems safe to say that a general forecast for global climate conditions decades out would be even less reliable.

Here is where meteorologists can play a positive role. If meteorologists united for the limited purpose of urging their sister science of climatology to return to realism and humility, then we could be spared a catastrophic loss of liberty and decline in wealth—regardless of what the climate does.

SideBar38: In Polite Company, It's Not Proper to Talk about Religion, Politics, *or* the Weather

Be careful what you say around the office water cooler. The president seems to be blaming everything these days on the weather...

There's an old saying that, in polite company, it's not proper to talk about religion or politics. It was considered safe to stick to neutral subjects like the weather to avoid conversational confrontation. Nowadays, the climate on that subject has changed dramatically.

In early 2015, I attended a presentation on recent and projected short-term weather conditions for southwestern Pennsylvania. The event was directed at specialists in the public-health profession. A meteorologist from the local National Weather Service office gave details on the region's recent cold snap and estimated when he expected a warm up for spring—all well and good, until an attendee turned the talk to anthropogenic climate change.

The participant was concerned that everything from local weather anomalies to ocean acidity and sea-level rise was attributable to people. The same ubiquitous, dubious soundbite arguments targeting the culpability of Americans living comfortably were tossed out as if they were indisputable facts. But the angst from the attendee was real.

Where did all this weather worry originate?

As a practitioner in the atmospheric science profession for 35 years, I have witnessed a great deal of change in the field of climatology. Gone are the days when climate science was focused on the tedious collecting, analyzing, and disseminating of facts and figures from local, regional, and global data. Today, it's easy to find fame and fortune in forecasting frightening futures.

Since the early 1980s, social and political opportunists have been riding the arrogant confidence of a handful of authoritative scientists to trample the objectivity of climate reality. So, a mundane chat that includes "the balmy temperatures we've been having lately," becomes "what are you doing to reduce your carbon footprint to stop this global warming?" The traditional "January thaw" morphs into clear evidence of "manmade climate change."

And now, many of our betters in the federal government assure us that apparently specially imbued climatologists can be trusted to tell us if the Earth will still be habitable a hundred years hence or, as a minimum, which energy sources to invest in.

Thus the divinations of a climate science high priesthood propped up by the power of the political class has warranted certainty about the future state of global conditions, which is certainly unwarranted given the reality that global average temperature has leveled off over the past 18 years. In fact, some scientists are seriously predicting an imminent substantial temperature drop worldwide based on trends in natural conditions that include sunspot activity and ocean circulation patterns.

Nevertheless, when a friendly conversation about the weather turns into a holy war against "climate deniers," it's time to recognize that the weather topic has become just another turbulent atmosphere suffused in religion and politics.

So be extra careful what you talk about—the list of subjects for polite discussion is shrinking.

A Tale of Two Forecasters

LATE IN 2010, progressives and environmental activists united for the UN's climate confab to save the planet in Cancun, Mexico. Yet just a few hundred miles northwest of Cancun in Galveston, Texas, at a time when humans were not thought to be distorting the climate, the deadliest natural disaster in US history occurred.

On September 8, 1900, the winds and waves of a fierce hurricane produced a sad and somewhat avoidable episode. The Weather Bureau's top official on Galveston Island that fateful September day was Dr. Isaac Monroe Cline. Dr. Cline's own recollections of the event are recorded in his *Storms, Floods and Sunshine: Isaac Monroe Cline, an Autobiography*.[106] He recalls harnessing his horse to the two-wheeled cart he normally used for hunting and driving it along the beach. Dr. Cline goes on:

> I warned the people that great danger threatened them, and advised some 6,000 persons...to go home immediately. I warned persons residing within three blocks of the beach to move to the higher portions of the city, that their houses would be undermined by the ebb and flow of the increasing storm tide and would be washed away. Summer visitors went home, and residents moved out in accordance with the advice given them. Some

6,000 lives were saved by my advice and warnings. (p. 92)

Today, the National Weather Service gives the Isaac M. Cline Awards to recognize the achievements and excellent work of its employees in saving lives and property.

Unfortunately, there is a problem. Dr. Cline's gallant efforts were likely a figment of his imagination.

In his national bestseller *Isaac's Storm: A Man, a Time, and the Deadliest Hurricane in History*,[107] Erik Larson describes a more likely scenario based on his meticulous investigation. According to scores of independent historical accounts, it appears that Dr. Cline was not a meteorological Paul Revere riding up and down the beach and urging bathers to seek shelter. Nor did he warn residents of impending doom until it was much too late for anyone to take the proper precautions. In fact, up until just a few hours before the brunt of the storm's fury hit Galveston, Dr. Cline staunchly reassured residents that it was nearly impossible for the island to suffer serious damage from a hurricane.

Larson's research flatly contradicts Cline's account of the events. It also raises questions about relying on the wisdom and heroism of government officials. An estimated 6,000 to 8,000 people, including Cline's pregnant wife, lost their lives during the storm.

Now compare the Isaac Cline affair with the life of another meteorology enthusiast and weather forecaster, Vice-Admiral Robert FitzRoy. In the early 1830s, Robert FitzRoy was captain of the *Beagle*, the ship that carried Charles Darwin to the Galapagos Islands and back, and played an important role in the development of the theory of evolution. Later in his career, FitzRoy pioneered modern weather forecasting. In 1854, he was selected to head the newly established meteorological department for England's Board of Trade. Though not as famous as Charles Darwin, FitzRoy was exceptionally devoted to his work, and amassed a long and impressive list of accomplishments and titles.

Unfortunately, Robert FitzRoy had unreasonably high expectations of himself. In the early years of national weather forecasting, accurate predictions were much more difficult to make than they are today, and inaccurate forecasts were quite common. To expect better results was not very realistic given the rudimentary equipment, sketchy observations, and incomplete theory of that time. Tragically, FitzRoy allowed his challenging and at times futile job to become his identity, until one morning in April of 1865, after repeatedly judging himself by unreasonable standards, the 59-year old FitzRoy took his own life.

Is there anything to be learned from the lives of these two historic figures that might temper attitudes about manmade global warming? One lesson is that arrogance can lead to calamitous loss of life. Though both men are to be admired for their intelligence, work ethic, and achievements, they thought too highly of themselves—though in distinctly different ways. Having too much confidence in his expert opinions, Dr. Cline dissuaded residents from fleeing inland and many paid with their lives. Placing excessive demands on himself, Mr. FitzRoy became despondent and committed suicide. Overconfidence and excessively high expectations of oneself are simply different forms of arrogance.

The biggest challenge confronting climate forecasting today is hubris. As it has in the past, hubris can lead to loss of life and economic disaster. Beware of scientists and government officials who believe it is possible to predict the future. Realize that their knowledge is woefully inadequate, though they won't admit it to you, because they probably haven't admitted it to themselves.

Over the coming months and years we will hear additional warnings about future catastrophes from people who earn their keep by touting the urgency as well as the certainty of such predictions. When considered in those terms, it's not surprising that they don't tolerate dissent.

There are many ways that a forecast can go wrong, and other

forecasts may only appear to be right. The ancients realized that a cleverly-worded prediction works regardless of the outcome. For example, King Croesus of Lydia asked the oracle at Delphi what would happen if he attacked Persia. The oracle responded, "If you do, you will destroy a great empire." But it was *his* empire that was destroyed.[108] The moderns have learned they can bamboozle people by making predictions about what will happen long after they are gone.

Forecasting is part science and part art. Many like Isaac Cline emphasize the part that makes their forecasts appear more authoritative—the science part. But forecasting involves a large dose of art and, therefore, is much less certain than some practitioners are willing to admit. As the forecaster looks farther into the future, it becomes more and more art, until finally it becomes nothing more than illusion. The art of long-range global climate forecasting, like any complex art, may be impressive to behold, but the forecasts themselves are too unimpressive to believe.

Restoring Integrity and Optimism to Science

THE POLITICIZATION OF science is bad for not only science, but education and government as well. Considerable damage has been and is being done, but it's not too late to turn things around. In this book, I've emphasized five key themes:

A scientist must be free to explore any hypothesis, theory, or doubt—Those who say only scientists who agree with the consensus view are "real scientists" have failed to learn the lessons of history. Great discoveries and theories are made by those who think outside the box—not by those who are trapped inside. The truth is not the winner of a popularity contest. Nor is science so fragile that it can't withstand criticism, vigorous debate, and fiercely independent research;

Good scientists practice humility – Good scientists understand that arrogance leads to errors. They recognize that humans are a relatively insignificant force in the universe. They are awed by the richness and mystery of nature;

Science must not be politicized – The purpose of science is to discover facts—not invent them. When politics and science are combined, science is corrupted. The work of experimentalists and

231

theoreticians must be free of political influences. Activists try to use science to promote their preconceptions and utopian goals. Scientists must pursue the truth wherever it leads;

Science literacy means understanding the difference between knowledge, on one hand, and assumptions, guesses, and beliefs, on the other hand. To the science establishment, science literacy means accepting the establishment's pet theories. These theories are treated as partially verified facts—despite the fact that scientific theories are by definition explanations. By confusing theory with fact, the science establishment is actually contributing to scientific illiteracy. For example, scientific illiteracy has enabled widespread deception about what is known and what can be reliably predicted about the Earth's climate; and,

Crisis-mongering is particularly harmful to climate science –
Today's crisis-mongering climate science is run like a shady business. It soaks the middle class and deprives the world's poor of a better tomorrow. Government funding is directed to topics that have been declared urgent. These topics attract researchers seeking funding. Conclusions are drawn that support the crisis-mongering and the need for further research.

Sadly, 21st century climate science is oppressed by ideas that should have been discarded long ago. Climate science is being crushed under the weight of the political baggage it's being forced to carry. Climate science is being held hostage by a top-down, centralized, authoritarian organization resembling the medieval church. Dogmatic science harms not only industry but education, government, civil discourse, and even science itself. There is too much at stake to allow this situation to continue.

Society and science will be better served when the current war between supporters and challengers of the anthropogenic global warming hypothesis yields to a more tolerant environment.

However, present trends suggest that this will not happen any time soon. More likely, climate conditions will continue to unfold in ways not predicted by the trusted climate models. When the public gets wind of this, they will put pressure on government to spend their tax dollars in more productive and beneficial ways such as understanding how to adapt to natural atmospheric variations. Rather than demanding a reduction in research funds, it may make more sense to demand research that is more diverse, more open to alternative hypotheses, and less self-congratulatory.

Ultimately, people (including scientists) believe what they want to believe. I hope that this book is not received as just another partisan broadside, but a plea to all of my colleagues in the fraternity of science to be more open-minded and more tolerant. There is one goal that we should all be able to agree on: to encourage and find rational solutions to demonstrable climate change and environmental challenges—solutions that benefit both people and the planet.

Finally, I realize that much of what's offered in this book may be considered heretical and unwelcome in the climate of today's science. But, as we all know, climate changes.

Acknowledgements

SPECIAL RECOGNITION AND gratitude goes to the following professionals who helped to make this book possible: Susan T. Cammarata, J.D., family and environmental lawyer in Pittsburgh, PA; JoAnn Truchan, P.E., chemical engineer; Stanley J. Penkala, Ph.D., chemical engineer and president of Air Science Consultants in Bridgeville, PA; and Albin Sadar, book author in New York City. These colleagues assisted in the writing of a few portions of some of the chapters. Albin Sadar, my identical twin brother, also helped edit the manuscript.

In addition, I am extremely grateful to my first edition publisher, Ira Brodsky of Telescope Books, for his encouragement and thoughtful contributions to the text. And am equally grateful for the opportunity to publish this revised and expanded version with Ken Coffman and Stairway Press.

Book chapters contain ideas and material first discussed in *The Washington Times*, *The Washington Examiner*, the *Pittsburgh Post-Gazette*, and the daily internet publication *American Thinker*.

Finally, this book represents my informed opinion after decades of work and reflection in the field of atmospheric science. I alone am responsible for any inaccurate statements that appear in this work. And though I criticize the ideas and actions of several well-known figures in the field of climate science, it is not my intention to attack or impugn them personally.

About the Author

ANTHONY J. SADAR is a Certified Consulting Meteorologist
with 35 years of experience in atmospheric and environmental
science and science education. During his career, he has divided
his time about equally among government, private industry, and
academia. Mr. Sadar is currently the air pollution meteorologist
and an air pollution program administrator for a large public
health department. In addition, he is an adjunct associate professor
of science and part-time instructor of meteorology and
climatology. He founded Environmental Science Communication,
LLC, a private consulting firm specializing in air pollution
dispersion modeling, regulatory compliance, and risk
communication for business and industry. His hands-on
experience includes weather observation above the Arctic Circle,
air quality modeling, and environmental project management.
Mr. Sadar has authored dozens of articles about atmospheric and
environmental issues.

Mr. Sadar's commentaries and book reviews have appeared in
The Washington Times, *The Washington Examiner*, and other
newspapers and trade publications. Mr. Sadar, along with Mark
Shull, authored *Environmental Risk Communication: Principles and
Practices for Industry* (CRC Press/Lewis Publishers, 2000).

He holds a BS in meteorology from The Pennsylvania State
University, an MS in environmental science from the University
of Cincinnati, and an MEd in science education from the

University of Pittsburgh.

In addition, he is a member of the American Meteorological Society; the Air & Waste Management Association; Kappa Delta Pi (education honors fraternity); and the Local Emergency Planning Committee of Allegheny County (Pittsburgh, PA).

Notes

FOR THE DOZENS of commentaries by Anthony J. Sadar and co-authors first published in *The Washington Times*, *The Washington Examiner*, and *AmericanThinker.com*, please access the individual websites.

Recommended Reading

FOLLOWING IS A very short list of several recent books that can provide a quick perspective on the climate science issue. I have used information from most of these books (plus numerous others, of course, as noted throughout *In Global Warming We Trust*) to inform my own thinking on this issue. In the list, "C" indicates that the author challenges the establishment position that humans are largely responsible for global climate change, while "E" indicates that the author endorses the establishment position that claims anthropogenic culpability. Each of these representative books is quite readable and ably explains and defends their respective positions.

Climate Coup: Global Warming's Invasion of Our Government and Our Lives edited by Patrick J. Michaels, 2011, Cato Institute — C

The Deniers: The World Renowned Scientists Who Stood Up Against Global Warming Hysteria, Political Persecution, and Fraud, and Those

Who are too Fearful To Do So by Lawrence Solomon, 2008, Richard Vigilante Books — C

Eco-Tyranny: How the Left's Green Agenda Will Dismantle America by Brian Sussman, 2012, WND Books — C

Environmentalism Gone Mad: How a Sierra Club Activist and Senior EPA Analyst Discovered a Radical Green Energy Fantasy by Alan Carlin, 2015, Stairway Press — C

The Great Global Warming Blunder: How Mother Nature Fooled the World's Top Climate Scientists by Roy W. Spencer, 2010, Encounter Books — C

Into the Dustbin: Rajendra Pachauri, the Climate Report & the Nobel Peace Prize by Donna Laframboise, 2011, Ivy Avenue Press — C

Climate Cover-Up: The Crusade to Deny Global Warming by James Hoggan with Richard Littlemore, 2009, Greystone Books — E

The Forgiving Air: Understanding Environmental Change, 2nd Revised Edition, by Richard C. J. Somerville, 2008, American Meteorological Society — E

The Hockey Stick and the Climate Wars: Dispatches from the Front Lines by Michael E. Mann, 2012, Columbia University Press — E

How to Change Minds about Our Changing Climate: Let Science Do the Talking the Next Time Someone Tries to Tell You... The Climate Isn't Changing, Global Warming is Actually a Good Thing, Climate Change is Natural, Not Man-Made... and Other Arguments It's Time to End for Good by Seth B. Darling and Douglas L. Sisterson, 2014, The Experiment, LLC —— E

Anthony J. Sadar

The Inquisition of Climate Science by James Lawrence Powell, 2011,
Columbia University Press — E

*Storms of My Grandchildren: The Truth about the Coming Climate
Catastrophe and Our Last Chance to Save Humanity* by James Hansen,
2009, Bloomsbury USA — E

Endnotes

---

[1] Renewable energy sources bring their own environmental headaches. See, for example, *New Scientist*, "How Clean is Green," by Anil Ananthaswamy, January 28/February 3, 2012, pages 34 - 38.

[2] "EPA already has in place rules requiring GHG [greenhouse gas] limits in some air permits, and continues to develop new GHG rules, including pending new source performance standards (NSPS) to cut GHG emissions from refineries and power plants" (from "EPA Database Could Pressure GHG Cuts, Serve As Basis For Air Law Suits," *Inside EPA*, "Clean Air Report," January 19, 2012, page 30). See also, "EPA Air Rules Head to Court" by Brent Kendall, *The Wall Street Journal*, February 27, 2012, page A2.

[3] *The Long Thaw: How Humans are Changing the Next 100,000 Years of Earth's Climate*, David Archer, (2009, Princeton University Press) page 158.

[4] See "Scientists Behaving Badly" by Rick Rinehart in *American Thinker*, February 25, 2012, for a former science editor's view of climate science practices today versus climate science in the 1970s and '80s.

[5] I am told that when Italian entrepreneur Guglielmo Marconi arrived in New York City after his historic transatlantic wireless transmission, one reporter asked him whether his giant spark transmitters were causing an increase in severe thunderstorms.

[6] See *Censoring Science: Inside the Political Attack on Dr. James Hansen and the Truth of Global Warming* (2008, Dutton) by Mark Bowen, page 223. "[Senator] Wirth and his staff... set the stage by leaving the windows of the hearing room open all that sweltering night to ensure that the air

conditioners would be working extra hard during the hearing the next day."

[7] See http://nofrakkingconsensus.com/climate-bible/

[8] Professor emeritus and atmospheric physicist Garth W. Paltridge writes in *The Climate Caper: Facts and Fallacies of Global Warming* (2010, Taylor Trade Publishing) that it can be difficult today "for the ordinary scientist to question the official beliefs of the apparatchiks of global warming." He continues, "The IPCC was always going to be a lobby mechanism for a particular view of the climate change issue, so it is not too surprising that it has become more than a little messianic and tends to ignore contrary opinion. Certainly its [behavior] argues a belief in the old adage that the end justifies the means. But its most remarkable achievement is that it has introduced a sort of religious supportive [fervor] into the [behavior] of many of the scientists directly involved in its activity" (p. 9.)

[9] I suspect interest in the atmospheric science field develops in much the same ways as interest in other vocations involving the natural world. Early on, children acquire a curiosity and aptitude for understanding this particular aspect of nature. Perhaps like me, a youngster begins by reading up on the topic, making crude weather instruments along with purchasing more sophisticated equipment, and taking regular weather observations. As time goes by, they may join or start a weather club at school and set up weather stations at remote locations. Eventually they discover a desire to pursue the field as a profession and explore colleges that offer degrees in meteorology and/or atmospheric science. Upon completing a rigorous academic program involving advanced mathematics, physics, chemistry, and the like, plus courses on theoretical and applied atmospheric science, the graduate may pursue further education or embark on what will hopefully be a worthwhile and fulfilling career. (For another example of a childhood interest in weather leading to a career, see the account by the current President of the American Meteorological Society, Louis W. Uccellini, in the January 2012 *Bulletin of the American Meteorological Society*, Vol. 93, Iss. 1, pages 100-101.)

[10] "Conference on Air Quality Modeling," January 31, 2012, *HealthTechZone.com,*

http://www.healthtechzone.com/news/2012/01/31/6089116.htm. I likewise attended the 11th Modeling Conference in August 2015.

[11] See the July 1985 *Bulletin of the American Meteorological Society*, Vol. 66, Iss. 7 pages 786-787. Dr. Robock is now at Rutgers University in the Department of Environmental Sciences.

[12] Throughout this book I use the various terms that are commonly used (and usually derogatorily) to describe those atmospheric scientists who are not convinced by the evidence that human-generated carbon emissions are causing long-term, global climate change. However, a more congenial term has been proffered by the Executive Director of the American Meteorological Society (see second note in Chapter 21).

[13] "No Need to Panic About Global Warming," January 27, 2012, *Wall Street Journal*, page A15. The signers included: Claude Allegre, former director of the Institute for the Study of the Earth, University of Paris; William Happer, professor of physics, Princeton; William Kininmonth, former head of climate research at the Australian Bureau of Meteorology; Nir Shaviv, professor of astrophysics, Hebrew University, Jerusalem; and [Antonino] Zichichi, president of the World Federation of Scientists, Geneva. The day after the OpEd appeared, there were more than 1,600 comments and more than 1,100 "Tweets" online, demonstrating tremendous interest in the subject. A letter to the editor in opposition was published February 1 under the title "Check With Climate Scientists for View on Climate," authored by Dr. Kevin Trenberth of the National Center for Atmospheric Research and signed by dozens of climate scientists. Their point was essentially that if you want to know what the Earth's climate will be like in the future you need to ask the climate experts—much as you would consult a cardiologist rather than a dentist about a heart condition. However, letters in rebuttal published on February 7 under the heading "The Anthropogenic Climate-Change Debate Continues" had perceptive words for the 38 experts. One writer, Peter Wilson, noted that though you would not consult a dentist for heart surgery, you might seek advice from an "intelligent generalist." After all, specialists such as heart surgeons (and by analogy climate scientists), "are easily biased by their specialization. When you're a hammer, everything looks like a nail." Another writer, Thomas H. Lauer, astutely observed that Dr.

Trenberth "gives the game away when he pronounces that a transition to a low-carbon economy will drive decades of economic growth. Only climate scientists are qualified to opine on climate, but somehow they are also qualified to explain global economics and political strategy."

[14] "Malaria, Politics and DDT," May 26, 2009, *Wall Street Journal*, http://online.wsj.com/article/SB124303288779048569.html

[15] While the amount of CO_2 has increased since the start of the industrial revolution, and its impact on climate is greater than its share of the atmosphere might suggest, it's still just 0.04% of the atmosphere's composition.

[16] *The Assault on Reason*, by Albert Gore (2008, Penquin)

[17] *Climate Cover-Up: The Crusade to Deny Global Warming* (2009, Greystone Books) by James Hoggan with Richard Littlemore.

[18] See the short video at:
http://www.youtube.com/watch?v=1yeA_kHHLow

[19] See the authors' bios at: http://www.desmogblog.com/about

[20] "The Royal Meteorological Society's statement on the Inter-Governmental Panel on Climate Change's Fourth Assessment Report," Professor Paul Hardaker, February 14, 2007, Royal Meteorological Society, http://www.rmets.org/news/detail.php?ID=332

[21] http://desmogblog.com/global-warming-denier-database

[22] *The Deniers: The World Renowned Scientists Who Stood Up Against Global Warming Hysteria, Political Persecution, and Fraud* (2008, Richard Vigilante Books) by Lawrence Solomon.

[23] See "Resistance to Warmism must be 'treated'," Thomas Lifson, March 31, 2012, *American Thinker*,
http://www.americanthinker.com/blog/2012/03/resistance_to_war mism_must_be_treated.html

[24] "The Question of Global Warming," Freeman Dyson, June 12, 2008, *New York Review of Books*,
http://www.nybooks.com/articles/archives/2008/jun/12/the-question-of-global-warming/?pagination=false. Dyson also discusses his "heretical thoughts" about global warming at:
http://www.edge.org/3rd_culture/dysonf07/dysonf07_index.html

[25] I recommend searching this website using the keywords "Climate of 2011."

[26] *The Climate Caper: Facts and Fallacies of Global Warming* (2010, Taylor Trade Publishing) by Garth W. Paltridge, page 17.

[27] From the lyrics to *Reelin' In the Years* by Steely Dan in his *Can't Buy A Thrill* album (ABC records, 1972).

[28] *The Wit and Wisdom of Oscar Wilde* (1999, Gramercy) by Oscar Wilde

[29] The idea that adults have more to learn from children than the other way around has become a popular theme in Hollywood.

[30] I also recommend Brian Sussman's more recent book, *Eco-Tyranny: How the Left's Green Agenda Will Dismantle America* (2012, WND Books) for additional insights about how Marxist ideology threatens liberty.

[31] For more of Dr. Deming's views on education see "What I Learned From a Brainiac," *Wall Street Journal*, "Opinion," page A15, February 1, 2012. See also his *Black & White: Politically Incorrect Essays on Politics, Culture, Science, Religion, Energy and Environment* (2011, CreateSpace).

[32] For a recent article by Dr. Singer, see "Peer Review Is Not What It's Cracked Up To Be," August 5, 2015,
http://www.americanthinker.com/articles/2015/08/peer_review_is_not_what_its_cracked_up_to_be.html

[33] https://www.theice.com/ccx.jhtml

[34] "Global status of DDT and its alternatives for use in vector control to prevent disease," by Henk van den Berg, Laboratory of Entomology, Wageningen University and Research Centre, Wageningen, the Netherlands, *Stockholm Convention on Persistent Organic Pollutants*, http://www.pops.int/documents/ddt/Global%20status%20of%20DDT%20SSC%20Oct08.pdf

[35] See "Marxism was, now isn't," *The Washington Times*, October 10, 2009 and *Intellectuals and Society* by Thomas Sowell (2009).

[36] *Nonsense on Stilts: How to Tell Science from Bunk* by Massimo Pigliucci (2010, University of Chicago Press)

[37] *Intellectuals and Society*, Thomas Sowell (2012, Basic Books)

[38] "Slapshot," *Wikipedia,* accessed on March 11, 2012 at http://en.wikipedia.org/wiki/Slapshot.

[39] *The Hockey Stick and the Climate Wars: Dispatches from the Front Lines* by Michael E. Mann (2012, Columbia University Press). For a more in-depth review with somewhat different conclusions see "A detailed review of Mann's book: *The Hockey Stick and the Climate Wars* as it relates

to the Wegman report to Congress" by Brandon Shollenberger, posted on March 7, 2012 by Anthony Watts at:
http://wattsupwiththat.com/2012/03/07/a-detailed-review-of-manns-book-the-hockey-stick-and-the-climate-wars-as-it-relates-to-the-wegman-report-to-congress/

[40] Or as one person commenting on my review (originally published in *The Washington Times*) of Dr. Mann's book suggested, the team might be better described as "Storm Troopers."

[41] *Climatology: An Atmospheric Science*, Third Edition, by John J. Hidore, John E. Oliver, Mary Snow, and Rich Snow (2010, Prentice Hall).

[42] Compare Figure 10.5, page 187, in the third edition of *Climatology: An Atmospheric Science* (corresponding to the first figure in this chapter) with the now discarded Figures 14.10 and 14.11, page 276, in the second edition (corresponding to the second figure in this chapter). Figure 10.5 was also included in a slightly different form as Plate 8 in the second edition.

[43] Because some tree ring proxy data did not reveal the same increase in air temperature from about 1960 to 1980 as thermometers, the thermometer measurements were used instead. Thus the phrase "hide the decline" that appeared in the Climategate e-mail messages.

[44] For example, in quantum mechanics the Heisenberg uncertainty principle says that it's impossible to know both the exact position and the exact momentum of a subatomic particle. This is just one of several arguments against determinism.

[45] A survey of members of the American Meteorological Society (AMS) on climate change science and perceptions was conducted by researchers at George Mason University's Center for Climate Change Communication. The survey results are discussed elsewhere in this book.

[46]
http://www.ucsusa.org/global_warming/science_and_impacts/science/ipcc-backgrounder.html#IPCC_Structure

[47] http://ipcc.ch/organization/organization_procedures.shtml

[48] "Climategate: the final nail in the coffin of 'Anthropogenic Global Warming'?" James Delingpole, November 20, 2009, *UK Telegraph*, http://blogs.telegraph.co.uk/news/jamesdelingpole/100017393/clim

ategate-the-final-nail-in-the-coffin-of-anthropogenic-global-warming/
[49]

http://blogs.scientificamerican.com/observations/2012/03/17/effecti
ve-world-government-will-still-be-needed-to-stave-off-climate-
catastrophe/
[50] http://www.sciencemag.org/content/335/6074/1306.summary

[51] "What is Science?" Richard Feynman, *The Physics Teacher,* Vol. 7 issue
6, 1969, pp. 313-320.
[52] For a longer list see:
http://blog.heritage.org/2009/11/17/global-warming-ate-my-
homework-100-things-blamed-on-global-warming/
[53] See http://www.ecoenquirer.com/global-warming-asteroid.htm
[54] See this page at the National Wildlife Federation website:
http://www.nwf.org/Global-Warming/Effects-on-Wildlife-and-
Habitat/Great-Lakes.aspx
[55] See the agenda and list of participants for the 6th International
Conference on Climate Change, June 30, 2011 in Washington, D.C.
[56] See:
http://www.americanprogress.org/pressroom/advisories/2011/06/cl
imate_denial_call
[57] See ExxonMobil's 2010 Worldwide Giving Report at:
http://exxonmobil.com/Corporate/community_wwgiving_report.aspx
[58] See:
http://www.gallup.com/poll/146810/Water-Issues-Worry-
Americans-Global-Warming-Least.aspx
[59] *Climate of Corruption: Politics and Power Behind the Global Warming Hoax*
by Larry Bell (2011, Greenleaf Book Group).
[60] *A Universe from Nothing: Why There Is Something Rather than Nothing*, by
Lawrence M. Krauss (2012, Free Press).
[61] Regarding the use of the word "unconvinced" to describe those who
are skeptical about the evidence for human produced climate change see
"Letter from Headquarters" in the December 2011 *Bulletin of the
American Meteorological Society* by that organization's Executive Director,
Keith L. Seitter. His refreshingly dispassionate article is titled "Dealing
Honestly with Uncertainties in Our Understanding of Climate Change."

In fact, that issue of the *Bulletin* focuses on uncertainty in atmospheric forecasts and includes an exceptional piece by Judith A. Curry and Peter J. Webster of the Georgia Institute of Technology titled "Climate Science and the Uncertainty Monster." Plus, the January 2012 *Bulletin* contains a perspective piece on "Calibration Strategies: A Source of Additional Uncertainty in Climate Change Projections," by Chun Kit Ho, et al.

[62] See ASTM International Standard D6589-05, 2010 (Section 11, Vol. 11.07, "Atmospheric Analysis," D22, Air Quality) on evaluating atmospheric dispersion model performance.

[63] *Slycraft's Catalog of Stuff* (1984, Crown Publishers) by Albin Sadar and Robert Pagani.

[64] For example, see *Science News*, "Engineering a cooler Earth" by Erika Engelhaupt, June 5, 2010.

[65] See "Get paid for eating" by Stanley J. Penkala and Anthony J. Sadar in the *Pittsburgh Post-Gazette*, April 5, 2009, starting on page G1.

[66] See "Favoring climate over weather," Rep. Ralph M. Hall, April 13, 2012, *Washington Times*, for a perspective on spending priorities. http://www.washingtontimes.com/news/2012/apr/13/favoring-climate-over-weather/

[67] For example, see *Human Nature: A Blueprint for Managing the Earth - by People, for People* (2004, Times Books) by the prolific science writer James Trefil.

[68] Though the Judeo-Christian faith has been around for thousands of years, so has environmentalism. Nature worship and animism are forms of environmentalism, as is the more recent speciesism. ("Speciesism refers to a bias against nonhuman animals simply because they are members of another species" says Lisa H. Sideris in her book *Environmental Ethics, Ecological Theology, and Natural Selection* (2003, Columbia University Press) page 136.) While ancient and modern environmentalism speak to the relationship between humans and the natural world, the same can be said about the Judeo-Christian faith. However, progressives might argue that only environmentalists look at that relationship from a rational and universal perspective. To the contrary, see *The Genesis of Science: How the Christian Middle Ages Launched the Scientific Revolution* by James Hannam (2011, Regnery Publishing)

and *The Victory of Reason: How Christianity Led to Freedom, Capitalism, and Western Success* by Rodney Stark (2006, Random House). Among other topics, *The Victory of Reason* describes Christianity's acceptance of diverse ideas and facts and logic, and its integral role in the rise of democratic, civil societies as well as great universities. (Even today, at the back of the stage in Harvard's Memorial Hall/Sanders Theater, the motto "Christo Et Ecclesia"—Christ and the Assembly (or Church)—can be seen around the familiar shield that bears the proclamation "Veritas"— Truth.)

[69] "Prices Soar on Crop Woes," Scott Kilman and Liam Pleven, January 13, 2011, *Wall Street Journal*, http://online.wsj.com/article/SB100014240527487048036045760777 51817700340.html

[70] For example, the Jewish philosopher Maimonides recognized that beliefs must not contradict demonstrable facts.

[71] For example, the Christian philosopher Thomas Aquinas believed that all reasonable arguments and counter-arguments deserved careful analysis.

[72] See *Human Nature: A Blueprint for Managing the Earth—by People, for People* (2004, Times Books) by James Trefil.

[73] The 2010 revenue figures for several leading environmental organizations are presented in "Fakegate: the smog blog exposes irrational rage, innumeracy, and heartland's efficient success" at joannenova.com, February 2012. Also see "The Not-So-Vast Conspiracy" by the editors of *The Wall Street Journal*, February 21, 2012, page A18, and "Global warming's desperate caper" by the editors of The Washington Times, February 24, 2012, page B2. For the flipside see "Green campaigns killed thousands of jobs in 2011" by George Landrith, *The Washington Examiner*, January 20, 2012, page 35.

[74] The first wind turbines used to generate electricity were deployed in 1887. See: http://en.wikipedia.org/wiki/History_of_wind_power#19th_century

[75] See http://en.wikipedia.org/wiki/AEP_v._Connecticut

[76] Dr. Michaels was one of the first, most persistent, and most prolific challengers to the anthropogenic global warming hypothesis. He has contributed numerous books and editorials. One of his recent OpEds,

"Is Global Warming a Bipolar Disorder?" (January 5, 2012), can be found at:
http://www.forbes.com/sites/patrickmichaels/2012/01/05/is-global-warming-a-bipolar-disorder/
This commentary, which includes three detailed maps, discusses satellite temperature measurements for the lower atmosphere gathered over the past 33 years. As Dr. Michaels notes, the data show that the latitudes northward of 60 degrees N warmed substantially while those southward of 60 degrees S cooled substantially. The rate of warming across the globe as a whole was about 40 percent lower than predicted by the UN's mid-range climate models. In another commentary ("A Sustainable Depression," January 9, 2012, *Washington Times*), Dr. Michaels addressed another important aspect of the carbon debate: energy sources. He details the economic failure and disruption that solar and wind energy have inflicted on the citizens of Germany, Spain, and the UK.

[77] *Climate Coup* was edited by Dr. Michaels and published in 2011 by the Cato Institute, a libertarian public policy research foundation located in Washington, DC

[78] "The Boscombe Valley Mystery," Sir Arthur Conan Doyle, *The Complete Sherlock Holmes* (1986, Doubleday: Bantam Classics).

[79] Sferics: the atmospheric discharges associated with storms.

[80] See *New Scientist*, "Luddite and proud," Interview by Alison George, December 24/31, 2011, pages 40 - 41.

[81] "Osama bin Laden lends unwelcome support in fight against climate change," Suzanne Goldenberg, January 20, 2010, *UK Guardian*, http://www.guardian.co.uk/environment/2010/jan/29/osama-bin-laden-climate-change

[82] ""Monkey bill" passes Tennessee Senate," March 20, 2012, National Center for Science Education, http://ncse.com/news/2012/03/monkey-bill-passes-tennessee-senate-007264

[83] See http://www.interacademies.net/

[84] As Michael Crichton noted in *State of Fear* (2004, Harper), "Environmental science is a contentious and intensely politicized field."

[85] In addition to the Unabomber, Earth First! Worldwide and the Earth

Liberation Front have practiced sabotage in the name of the environment. Acts of "ecotage" have included arson causing tens of millions of dollars worth of damage.

[86] "Lysenkoism," Robert T. Carroll, *The Skeptic's Dictionary* (2003, John Wiley & Sons). Under the direction of Trofim Denisovich Lysenko in the Soviet Union in the mid-1900s, "science was guided not by the most likely theories, backed by appropriately controlled experiments, but by the desired ideology. Science was practiced in the service of the State, or more precisely, in the service of ideology." www.skepdic.com/lysenko.html, "Lysenkoism")

[87] An opinion piece reflecting further on the intrusion of ideology into climate science by Daniel Greenfield titled "The Global Warming Cult and the Death of Science" appeared in the February 20, 2012 edition of *FrontPage Magazine* and can be viewed at: http://frontpagemag.com/2012/02/20/the-global-warming-cult-and-the-death-of-science/

[88] For an excellent discussion of what could happen to the US if it embraces European-style centralization of power, read *The New Road to Serfdom: A Letter of Warning to America* (2010, HarperCollins) by Daniel Hannan, a conservative British Member of the European Parliament.

[89] From *The Hockey Stick and the Climate Wars: Dispatches from the Front Lines* by Michael E. Mann (2012, Columbia University Press), page 75.

[90] "Right-wingers are less intelligent than left wingers, says study," Rob Waugh, February 8, 2012, *UK Daily Mail*, http://www.dailymail.co.uk/sciencetech/article-2095549/Right-wingers-intelligent-left-wingers-says-controversial-study--conservative-politics-lead-people-racist.html

[91] For example, leftists are trying to silence on-air meteorologists who don't subscribe to AGW. In early 2012, ThinkProgress Green launched its "Forecast the Facts" campaign to pressure the American Meteorological Society to strengthen its public statement about climate change and to discipline TV meteorologists who dared to express dissenting views. It is understandable that many TV weathercasters are skeptical about long-term global climate prognostications. After all, they are reminded daily of the difficulty of making long-range and even short-range predictions. They have earned the right to be skeptical

because their predictions can at least be tested on a regular basis.

[92] My sense is that the majority of people who question the idea that humans are substantially responsible for long-term global climate change do not doubt that significant global warming has occurred; however, they think that the warming is overwhelmingly from natural causes.

[93] There is a great deal of similarity between books such as *The Inquisition of Climate Science* and *Climate Cover-Up: The Crusade to Deny Global Warming*. Plus, Dr. Powell voices many of the same complaints and engages in the same name-calling as Dr. Michael Mann does in his book, *The Hockey Stick and the Climate Wars: Dispatches from the Front Lines*. Obviously, these books draw from many of the same partisan sources.

[94] See http://www.foxnews.com/scitech/2010/02/11/obama-spending-increase-global-warming-research/

[95] See http://cnsnews.com/news/article/obama-requests-770-million-fight-global-warming-overseas

[96] See http://www.eci.ox.ac.uk/publications/downloads/boykoff04-gec.pdf. For an update on Maxwell Boykoff's research, see *Who Speaks for the Climate?: Making Sense of Media Reporting on Climate Change* (Cambridge University Press, 2011).

[97] *Environmental Risk Communication: Principles and Practices for Industry*, by Anthony Sadar and Mark Shull, (CRC Press/Lewis Publishers, 2000).

[98] An informed opinion about the environmental and economic aspects of the tar sands and pipeline project is provided in "Voodoo environomics: Fantasy replaces reality in Obama's green economy" by geologist H. Leighton Steward, *The Washington Times*, February 20, 2012, page B1.

[99] Fracking (short for hydraulic fracturing) extracts natural gas by injecting a mix of chemically-treated water and sand under high pressure into shale formations. The technique has been applied recently in deep wells (thousands of feet underground) such as the Marcellus Shale formation in West Virginia, Pennsylvania, and New York. To learn about the potential hazards, see "The Risks of Shale Gas Development: How RFF Is Identifying a Pathway toward Responsible Development" in *Resources*, No. 179, 2012, pages 16-18 (published by Resources for the Future, RFF). RFF is in an 18-month project "to identify the perceived and potential burdens that shale gas development might impose on the

environment and on the community, and to think about how industry and government might address those," according to Alan Krupnick, Director of RFF's Center for Energy Economics and Policy.

[100] Businesses must answer to their shareholders, employees, and customers. If they do wrong, they may be prosecuted. National governments are accountable to their electorate—assuming they have free and fair elections. The UN is only answerable to its member states, most of which are neither free nor well-governed.

[101] The United States Environmental Protection Agency (EPA) is responsible for ambient (outdoor) air. It's ominous that the EPA is using the threat of climate change as an excuse to meddle in household air quality issues. Regardless, even state agencies are concerned about the recent regulatory excesses of the EPA. For example, some states have sued the EPA over its new sulfur dioxide ambient air quality standard, asserting among other things that the standard is more stringent than required by the federal Clean Air Act. The new one-hour standard is making industry compliance difficult, to say the least. For a broader perspective on the EPA's actions, see *Regulators Gone Wild: How the EPA is Ruining American Industry* by Rich Trzupek (Encounter Books, 2011).

[102] *The Delinquent Teenager Who Was Mistaken for the World's Top Climate Expert* by Donna Laframboise (Ivy Avenue Press, 2011)

[103] Furthermore, in *Climate Coup* (2011), economist Ross McKitrick claims that "[u]nfortunately, the way the IPCC works, it is allowed to make stuff up; then it's the job of its critics to prove it wrong" (p. 85).

[104] "Obama Axes Ozone Rule; Could Have Cost Business $90bn," September 2, 2011, *Environmental Leader*, The President had proposed stricter Ozone National Ambient Air Quality Standards. However, in response to criticism from industry, he withdrew the proposal. http://www.environmentalleader.com/2011/09/02/obama-axes-ozone-rule-revamp/

[105] "EPA Database Could Pressure GHG Cuts, Serve As Basis For Air Law Suits," *Inside EPA*, "Clean Air Report," January 19, 2012, page 30. "EPA already has in place rules requiring GHG [greenhouse gas] limits in some air permits, and continues to develop new GHG rules, including pending new source performance standards (NSPS) to cut GHG emissions from refineries and power plants" See also, "EPA Air Rules

Head to Court" by Brent Kendall, *The Wall Street Journal*, February 27, 2012, page A2.

[106] *Storms, Floods and Sunshine: Isaac Monroe Cline, an Autobiography* (Pelican Publishing Company, 1945)

[107] *Isaac's Storm: A Man, a Time, and the Deadliest Hurricane in History* by Erik Larson (Vintage Books, 2000)

[108] That story is according to the ancient Greek historian, Herodotus. See http://www.delphic-oracle.info/delphic-oracle.asp

Index

air circulation patterns, 140
American Clean Energy and
    Security Act, 152, 161
*American Electric Power (AEP)
    v. Connecticut*, 165
American exceptionalism,
    127
American Geophysical
    Union
    AGU, 205
American Meteorological
    Society
    AMS, 25, 37, 39, 43, 71,
        115, 174, 206, 236
anthropogenic global
    warming, 3, 5, 8, 31, 32,
    37, 52, 63, 64, 70, 71,
    77, 88, 99, 118, 123,
    155, 179, 180, 208, 217,
    232
Archimedes, 1
Arctic Circle, i, ii, iii, 185,
    235
artificial trees, 135
Asteroid, 124
ASTM International, 132
atmospheric science, i, 7,
    17, 18, 22, 23, 26, 33,
    38, 39, 43, 48, 64, 78,
    84, 126, 155, 169, 181,
    196, 202, 203, 219, 225,
    234
aurora borealis, ii, 185
Barton
    David, 75
Beacon Power, 2
Bell
    Larry, 107, 108, 126,
        127
bin Laden
    Osama, 179, 180
Boykoff
    Jules M., 209
    Maxwell, 209
BP
    British Petroleum, 190
Browner
    Carol, 116
Bush
    George W., 77
cap and trade, 63, 152
Cape Lisburne
    Alaska, i, ii
carbon credits
    offsets, 5, 46, 128, 218
carbon dioxide
    CO_2, v, 8, 14, 19, 30,
        35, 46, 47, 55, 62, 69,
        89, 93, 97, 103, 105,
        108, 112, 115, 122,

CPSIA information can be obtained
at www.ICGtesting.com
Printed in the USA
FSHW020646141119
63994FS